# Dryland Vegetation in the Sand-Shifting Environment

Ken Yoshikawa • Norikazu Yamanaka
Nobuhito Ohte • Naoko H. Miki • Linhe Wang
Guosheng Zhang

# Dryland Vegetation in the Sand-Shifting Environment

## Mu Us Sandy Land, China

Springer

Ken Yoshikawa
Faculty of Environmental, Life,
Natural Science and Technology
Okayama University
Okayama, Japan

Nobuhito Ohte
Biosphere Informatics Laboratory,
Graduate School of Informatics
Kyoto University
Kyoto, Japan

Linhe Wang
College of Forestry
Inner Mongolia Agricultural University
Hohhot, Inner Mongolia, China

Norikazu Yamanaka
Arid Land Research Center
Tottori University
Tottori, Japan

Naoko H. Miki
Faculty of Environmental, Life,
Natural Science and Technology
Okayama University
Okayama, Japan

Guosheng Zhang
College of Forestry
Inner Mongolia Agricultural University
Hohhot, Inner Mongolia, China

ISBN 978-981-95-4734-0 ISBN 978-981-95-4735-7 (eBook)
https://doi.org/10.1007/978-981-95-4735-7

The original submitted manuscript has been translated into English. The translation was done using artificial intelligence. A subsequent revision was performed by the author(s) to further refine the work and to ensure that the translation is appropriate concerning content and scientific correctness. It may, however, read stylistically different from a conventional translation.



This Springer imprint is published by the registered company Springer Nature Singapore Pte Ltd.
The registered company address is: 152 Beach Road, #21-01/04 Gateway East, Singapore 189721, Singapore

# Introduction

Rapid global warming and population growth are disrupting, degrading, destroying, and eliminating ecosystems all over the planet. Particularly in drylands at the limits of plant growth, slight changes in environmental conditions can lead to significant environmental degradation, causing almost irreversible land degradation known as desertification.

Therefore, in order to preserve the global environment and maintain people's living areas, it is essential to expand knowledge related to the characteristics of ecosystems in marginal areas and their protection and restoration.

This book aims to explain the characteristics of dryland ecosystems, focusing on the Mu Us Sandy Land, which spans over 30,000 km$^2$ in the central part of the Ordos Plateau in northern China (Fig. 1). Despite the cold and dry conditions of the Mu Us Sandy Land and its predominance of shifting sands, which create an extremely stressful environment for flora and fauna, surprisingly diverse ecosystems have been established. Among them, the unique community dynamics and ecophysiological characteristics of the evergreen conifer *Juniperus sabina* (=*Sabina vulgaris*), which can form dense communities in sandy areas, demonstrate the high adaptability in drylands.

Since 1985, experts in arid ecology and grassland physiology from the Inner Mongolia Autonomous Region of China, along with Japanese experts in forest ecology, tree physiology, and hydrology, have been conducting joint research to elucidate the mechanisms of desertification and develop greening technologies as countermeasures against desertification in drylands of Central Asia. Forty years have passed since the start of the research, and the authors of this book have spent most of their academic careers in the Mu Us Sandy Land, producing many research results, mainly on *J. sabina*. The purpose of this book is to compile these results into one volume to serve as a milestone for future generations and a springboard for further research on drylands.

This book is composed of six parts. Part I provides an overview of the origin of the Mu Us Sandy Land and its current state of land degradation, and presents the environmental characteristics of the drylands covered in this book. Part II reveals the site characteristics of the shifting sand environment in the Mu Us Sandy Land,

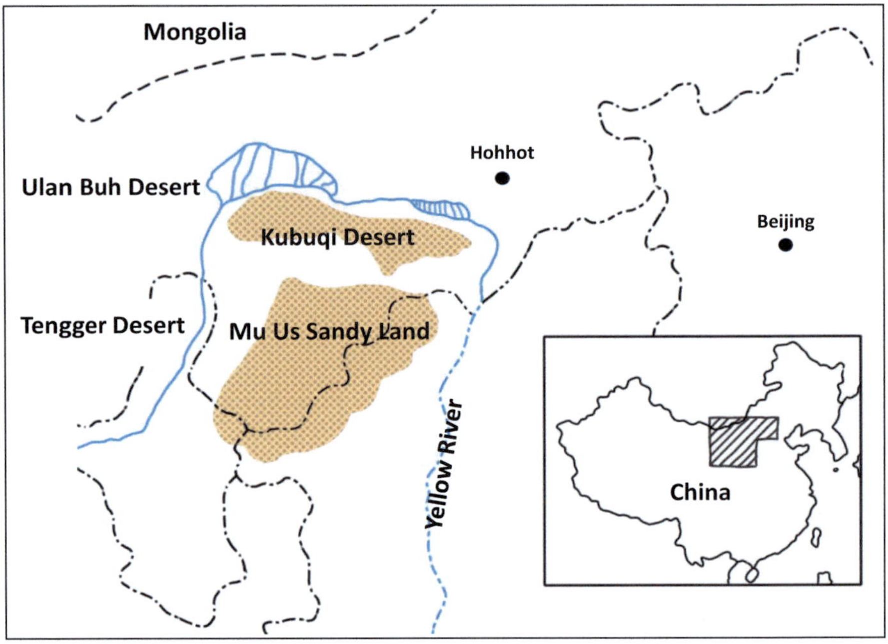


**Fig. 1** Location of the Mu Us Sandy Land

such as environmental conditions and topography as a dryland, and explains the characteristics of plant communities adapted to it, demonstrating the richness of dryland ecosystems. Part III explains the ecophysiological characteristics that plants have acquired to grow in cold and dry sites, from the perspective of rapid drying stress and high light stress tolerance, and shows the tenacious tactics of plants in the marginal area of their distribution. Part IV focuses on *J. sabina*, which is adapted to the cold and dry shifting sand environment of the Mu Us Sandy Land, and willows used for greening in drylands, and explains their unique life history strategies and tree species characteristics, demonstrating the robustness and delicacy of xerophytes in adapting to the environment. Part V explains the current state of desertification and its background that drylands, including the Mu Us Sandy Land, are facing and attempts to provide a future outlook for the restoration of arid ecosystem. Finally, Part VI describes the technologies that have been developed to combat desertification in China over a long period of time and provides an outlook on future greening technologies.

The scientific names of plants in this book are described according to the Flora of China (http://www.efloras.org/) and WFO Plant list (https://wfoplantlist.org/).

# Competing Interests

The authors have no competing interests to declare that are relevant to the content of this manuscript.

# Contents

# Part I
# Mu Us Sandy Land

The Mu Us Sandy Land, which this book takes up as a case study area, is a region where cold weather strongly limits the possibility of life, even among drylands. Also, the site conditions are dominated by sandy soils, which are subject to strong wind erosion, creating a unique landscape of shifting sand dunes. Desertification/land degradation progressing in such cold, dry, and shifting sand environments is one of the main land degradation phenomena among the various types of desertification phenomena occurring in drylands worldwide. After explaining the definitions of drylands and desertification/land degradation, we will look at the current state of desertification in China and provide an overview of the origin and current situation of the Mu Us Sandy Land.

# Chapter 1
# Drylands and Desertification/Land Degradation

**Ken Yoshikawa**

## 1.1 What Is Dryland

Dryland is defined as an area where the aridity index (AI), which represents the degree of dryness (AI = P (annual precipitation)/PET (annual potential evapotranspiration)), is 0.65 or less. These areas are divided into four categories based on the value of AI: hyper-arid areas $<0.05 \leq$ arid areas $<0.20 \leq$ semi-arid areas $<0.50 \leq$ dry sub-humid areas $<0.65$. Hyper-arid areas with an AI value of less than 0.05 are mostly deserts where plants cannot grow and are therefore excluded from the scope of "desertification" as defined in UNCCD (the United Nations Convention to Combat Desertification).

The distribution of drylands is around the tropical convergence zone of the north and south tropics, on the leeward side of large mountain ranges, in the interior of continents far from oceans, and along the coastal areas affected by cold currents flowing from the poles toward the equator. In the center of these drylands, there are mostly hyper-arid areas, and the degree of aridity decreases gradually toward the periphery.

The estimated area of dryland varies widely depending on the calculation method and data used, but many studies estimate it to be more than 40% of the world's land. As global warming progresses, the area of dryland is expected to increase further, possibly by about 7% more than now by the end of this century (Fig. 1.1).

## 1.2 What Is Desertification/Land Degradation?

The vegetation of the Mu Us Sandy Land has established itself under extremely harsh environment for plant growth. Moreover, recent global warming has made the environment of drylands much more severe than before.



K. Yoshikawa et al., *Dryland Vegetation in the Sand-Shifting Environment*,
https://doi.org/10.1007/978-981-95-4735-7_1

**Fig. 1.1** Sand dunes covered in vibrant ripple marks. (The Badain Jaran Desert, Ejina, Inner Mongolia Autonomous Region)

Land degradation includes the lowering of groundwater levels, the reduction of surface water, the salinization of soil, the erosion of topsoil, and the devastation of native vegetation. Aubréville (1949) first described the progression of land degradation in drylands to a desert-like state as desertification. However, the phenomenon he captured was the transformation of tropical forests into savannah-like states (savannization) and further transformation of savannahs into desert-like states (desertification) due to human activity. He attributed land degradation to anthropogenic destruction rather than to climate aridity (Kadomura 1988), because at that time, no one considered that climate changes would cause large-scale and short-term shifts in ecosystems showing zonal structures at the global level. However, global warming due to climate change has become a reality, completely changing the situation from that time.

In other words, land can degrade not only due to changes in the intensity of human activity but also due to changes in natural factors, so land degradation can occur anywhere in the world. Land degradation in drylands is particularly referred to as "desertification," but desertification/land degradation has become significantly different from what Aubréville initially described.

Among the various definitions of desertification that have been proposed so far, the most recent definition of desertification is given in the Special Report on Climate Change and Land (SRCCL) (IPCC (Intergovernmental Panel on Climate Change) 2019). It is based on the IPCC's 1994 definition and is in line with the IPCC's Fifth Assessment Report (AR5) (IPCC 2013). In other words, desertification is land degradation in arid areas due to various factors including climate change and human activity (UNCCD Article 1) (Fig. 1.2).

This land degradation refers to the reduction or loss of biological or economic productivity and integrity due to processes or combinations of processes, including patterns of human activity and habitation, such as soil erosion caused by wind and water, physical, chemical, and biological or economic deterioration of soil properties, and long-term loss of natural vegetation in arid, semi-arid, and sub-humid areas, including rainfed farmlands, irrigated farmlands, ranges, pastures, forests, and woodlands (Veron et al. 2006).

The sentences in the legal text are difficult to understand, so to make it clearer, it would be as follows. Land degradation is the long-term decrease and disappearance of the productivity and health of the ecosystems of a certain land, and the value of a healthy ecosystem for the people living there. The main cause of land degradation, which threatens both the environment and human livelihoods, is human-induced climate change, but the way it affects varies from direct to indirect through a series of processes.

**Fig. 1.2** Schematic illustration on community degradation due to desertification. (**a**) Landscope before desertification. Rain falls, rivers flow, crops thrive. Firewood is harvested from nearby forests, and various livestock graze on the grasslands. Dunes lie far away. (**b**) Landscape after desertification has progressed. Rainfall has decreased, crops fail to yield, trees wither and die, and firewood must be transported from afar. Sand dunes encroach, and many livestock now live on the limited grasslands (Drawn by R. Yamamoto)

## 1.3 Is the Desert Expanding?

The concept of desertification/land degradation inherently includes a priori the idea that deserts are expanding. This is largely due to Lamprey's report in 1975 that the southern edge of the Sahara Desert expanded southward at an average annual rate of 5–6 km between 1958 and 1975 in North Darfur and North Kordofan in Sudan, moving the edge of the Sahara Desert by 90–100 km in 17 years (Kadomura 1988).

Despite much debate among experts about the expansion of deserts (Young et al. 2009), it has been adopted by organizations like UNEP (United Nations Environmental Programme) and FAO (Food and Agriculture Organization of the United Nations) since the 1970s. Moreover, the Sahel region had experienced two serious droughts in the mid-1970s and 1980s, and it was said that a spectacular expansion of the Sahara Desert would soon threaten all African arable land. Since then, measures such as afforestation, reduction of livestock numbers, prohibition of tree and herb cutting, and soil conservation have been implemented against desertification.

However, due to the lack of appropriate resources, the UNCOD (United Nations Conference on Desertification) (1977) and UNEP (1992) reports, which diagnose desertification based on observations at a certain moment and extrapolate them to create large-scale desertification maps, have been questioned for not considering seasonal cycle of vegetation and its inter-annual variation (Benjamin 2017). Therefore, Tucker and Nicholson (1999) used long-term satellite observations to test the hypothesis of the expansion of the Sahara Desert, showing that the "expansion" of the Sahara Desert actually depends on the inter-annual variability of precipitation. This means that Lamprey's results were just showing the changes in vegetation distribution caused by the year-to-year difference in precipitation, as they were based on two surveys in a wet year (1958) and a dry year (1975) and did not indicate the expansion of the Sahara Desert. Consequently, by the 1990s, it was considered that the Sahara Desert was not expanding southward, and now it is thought that the "expanding desert" is nothing more than a myth (Wu et al. 2008).

Still, without even mentioning the digital tattoos on social media in recent years, once information is disseminated to the public, it is hard to correct, and many people still believe that the desert in Africa is rapidly expanding. However, even if there is no desert expanding at a rate of 5–6 km per year as Lamprey reported, it does not mean that the desert is stable. Of course, it is now recognized that the edges of the desert move back and forth in close relation to climate change and seasonal and annual variations in precipitation. And as a result, if shifting sand dunes move from a desert to a non-desert area by wind, the desert is considered to have expanded (Fig. 1.3).

Reflecting these phenomena, even in 2007, UNEP concluded as follows. Since the 1930s, when records of precipitation and vegetation began, the boundary between the desert and semi-desert in the Sahel region has shifted 50–200 km southward and continues to move due to decreased precipitation. As a result, for example, there is a risk of further desertification in the sandy, dry savannahs and semi-deserts that account for 25% of Sudan's arable land.

**Fig. 1.3** In shifting sand areas, sand creeps across paved roads

## References

Aubréville A (1949) Climats, forêts et désertification de l'Afrique tropicale. Société d'éditions géographiques, maritimes et coloniales, Paris, p 351

Benjamin S (ed) (2017) Rural societies in the face if climatic and environmental changes in West Africa. Zotero. https://doi.org/10.4000/books.irdeditions.12298

IPCC (2013) Carbon and other biogeochemical cycles. In: Climate change 2013: the physical science basis. Contribution of Working Group I to the Fifth Assessment Report, pp 1370

IPCC (2019) Climate change and land. Special report on climate change, desertification, land degradation, sustainable land management, food security, and greenhouse gas fluxes in terrestrial ecosystems, p 896

Kadomura H (1988) Desertification studies: a historical review. Geogr Rev Jpn 61(2):205–228

Tucker CJ, Nicholson SE (1999) Variations in the size of the Sahara Desert from 1980 to 1997. Ambio 28:587–591

UNCOD (1977) Proceedings of the desertification conference (Nairobi, UNEP). Pergamon Press

UNEP (1992) World atlas of desertification, 1st edn. Arnold, p 69

Veron SR, Paruelo JM, Oesterheld M (2006) Assessing desertification. J Arid Envir 66:751–763

Wu W et al (2008) Land degradation monitoring in the west Muus, China. Int Arch Photogramm Remote Sens Spat Inf Sci 37:847–858

Young H et al (2009) Livelihoods, power, and choice: the vulnerability of the Northern Rizaygat, Darfur, Sudan. Tufts University, Feinstein International Center, Boston, p 115

# Chapter 2
# Desertification/Land Degradation in China

**Ken Yoshikawa**

## 2.1 Spread and Progress of Land Degradation

Due to the impacts of climate change, drylands have been increasing and are expected to continue to increase in the future. In China, a "National Survey on Desertification and Sandification (NSDS)" has been conducted every 5 years since 1994, and the sixth survey was compiled in 2019 over a period of 2 years (Fig. 2.1) (Zan et al. 2023). In the first survey from 1994 to 1996 (China National Committee for the Implementation of the UN Convention to Combat Desertification (CCICCD) 1997), 27.3% (2.26 million km$^2$) of China's land area was desertified. Of this, 44% was in arid areas, 35% in semi-arid areas, and 21% in dry sub-humid areas. Among the desertified lands, there were 77,000 km$^2$ of farmland, which accounted for 40% of the farmland. Also, 56.6% of the grasslands, or 1.05 million km$^2$, were desertified. The desertified lands included 1000 km$^2$ of forests. According to the sixth Report of NSDS (2019), the total area of desertified lands nationwide reached 2.57 million km$^2$, accounting for 26.8% of China's land area, and the area of severely and extremely desertified lands accounted for more than 34% of the total desertified land area nationwide. Looking at the causes, 1.81 million km$^2$ of land was desertified by wind erosion, 0.25 million km$^2$ by water erosion, 0.16 million km$^2$ by salinization, and 0.36 million km$^2$ by freeze-thaw. More than 70% of all desertified lands in China are due to wind erosion. As can be seen from Fig. 2.1, the total area of desertified lands nationwide decreased by 38,000 km$^2$ from 2.61 million km$^2$ in 2014–2019. However, only 100 km$^2$ of the desertified lands have been biologically protected, and there are still 0.28 million km$^2$ of land that is likely to be desertified in the boundary zones of agriculture and livestock farming nationwide, and if 0.19 million km$^2$ of grasslands, 0.05 million km$^2$ of farmland, and 0.03 million km$^2$ of forests are not properly protected, there is a high possibility of becoming new desertified lands.



K. Yoshikawa et al., *Dryland Vegetation in the Sand-Shifting Environment*,
https://doi.org/10.1007/978-981-95-4735-7_2

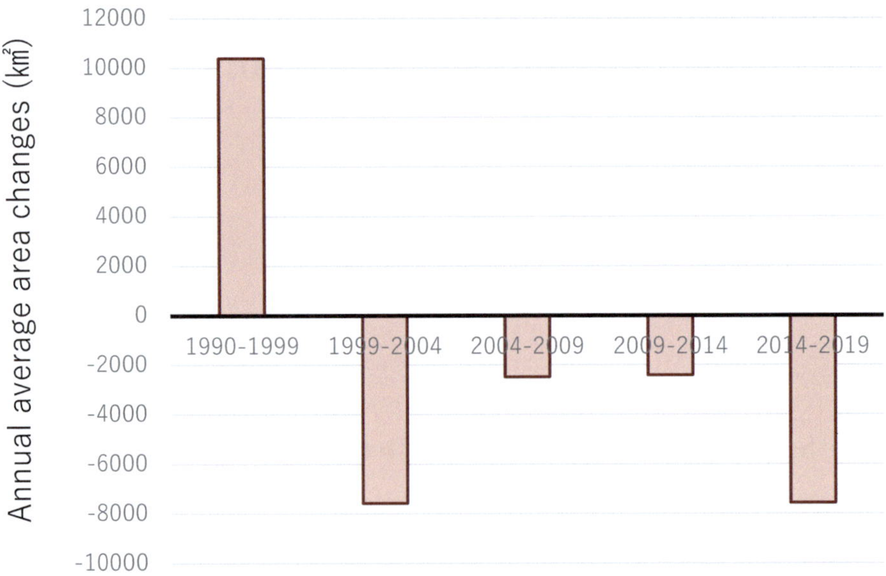


**Fig. 2.1** Annual average area changes of desertification land in China. (Redrawn from Zan et al. 2023)

## 2.2 Causes of Desertification/Land Degradation

The factors causing desertification/land degradation include both natural and human factors. Both types of factors not only degrade land but also act to improve land conditions. Furthermore, as these factors interact with each other, the effects of their actions become complex (Fig. 2.2).

Climate change affects temperature and precipitation, regulating the rate of photosynthesis, which in turn influences plant growth. Also, land degradation due to climate change ranges from concentrated events in a very short time, such as a 10-min storm removing topsoil or causing gullies and landslides, to slow nutrient depletion from the soil over a century, and loss of soil particles by wind and water, with a variety of time scales. Similarly, in human activities, changes in land use such as the land reclamation for agriculture and deforestation cause direct damage to dryland ecosystems in a short period. On the other hand, overuse of land, such as overgrazing and afforestation in grasslands, takes time to show their effects (Fig. 2.3).

As will be mentioned later, grasslands occupy a high proportion in the landscape of the Mu Us Sandy Land. China has 3.93 million km$^2$ of such grasslands, mainly used for livestock farming, accounting for 40% of the country's land area. In particular, there are five grasslands in the northwest and the Tibetan Plateau, accounting for 78% of all grasslands in China (Jiang et al. 2021). These areas are highly susceptible to the impacts of climate change and human activities due to their harsh

**Fig. 2.2** Reclaimed farmland on fixed sand dunes

**Fig. 2.3** Over-deforestation. The loss of dryland forests and subsequent land degradation occures through the extensive use of timber and firwood in drylands. (Illustrated by R. Yamamoto)

climatic conditions such as low rainfall, strong winds, cold, dryness, poor soils, and sparse vegetation. One-third of the grasslands in the northeast and Inner Mongolia has degraded over the past 40–50 years, and nearly 0.62 million $km^2$ of natural grasslands in Inner Mongolia has severely desertified in a few decades. Looking at the entire grassland, it is reported that 0.87 million $km^2$ has degraded due to climate change and excessive human activities associated with population growth (Fig. 2.4).

Furthermore, the pattern of desertification in China's arid and semi-arid regions differs east and west of the 100th meridian east (Wang et al. 2008). On the east side, it is greatly influenced by precipitation, wind speed, and other climatic factors. The west side is also influenced by the same factors, but in addition, terrestrial water (runoff from glaciers, rivers carrying meltwater, and the lakes into which they flow) plays a significant role in the progression and recovery of desertification. Therefore, on the west side, the development and decline of oases due to changes in river courses and surface water areas are closely related to regional land degradation. This is linked to the expansion of sandy areas due to the disappearance of vegetation and the vegetation restoration in surrounding areas such as the Ejin Banner Oasis (located at the 100th meridian) and the Taklimakan Desert, Tengger Desert, and Badain Jaran Desert (Fig. 2.5).

The continuous rise in temperature in recent years, as discussed later, has accelerated the melting of glaciers in the northwest, increasing the outflow of inland

**Fig. 2.4** Overgrazed grasslands. Land degradation bigins around watering places (Erenhot, Inner Mongolia Autonomous Region)

**Fig. 2.5** Grassland supported by glacier. Melting of the glaciers on the high mountains behind is progressing, causing a temporary increase in river flow (Xinjian Uygur Autonomous Region)

rivers in the short term. However, if the glaciers continue to melt, they will eventually disappear, resulting in a decrease in moisture and a different aspect of desertification in this region. Further research is needed on the long-term impacts of climate change and sustainability of ecosystems.

Climate change brings unprecedented dryness and high temperatures, reducing the stress tolerance of natural vegetation and ecosystems adapted to the original environment and increasing susceptibility to human activities. Moreover, it accelerates the current rate of land degradation, expands its scale, and brings new degradation patterns. Global warming so far has increased the frequency, intensity, and total rainfall of heavy rains and has advanced land degradation due to intense heat stress. As global warming progresses further, the frequency of floods and droughts will increase, and the severity of damage will increase, further worsening land degradation. Climate change gradually causes large-scale and long-term land degradation through a prolonged process. The global warming induced by human activities is a natural factor still changing its impact and is intensifying its severity.

## References

China National Committee for the Implementation of the UN Convention to Combat Desertification (CCICCD) (1997) China country paper to combat desertification. China Forestry Publishing House, p 36

Jiang P et al (2021) Climate and anthropogenic influences on the spatiotemporal change in degraded grassland in China. Environ Engineering Sci. https://doi.org/10.1089/ees.2020.0541
Wang X et al (2008) Desertification in China: an assessment. Earth Sci Rev 88:188–206
Zan GS et al (2023) Key data results and trend analysis of the sixth national survey on desertification and sandification. Forest Resources Management. https://doi.org/10.13466/j.cnki.lyzygl.2023.01.00

# Chapter 3
# The Mu Us Sandy Land and Its Origin

**Ken Yoshikawa**

## 3.1 What Is the Mu Us Sandy Land

The Mu Us Sandy Land is one of the four major sandy lands, namely Hulun Buir, Horqin, Hunshan Dake, and Mu Us, located mainly in the Ordos Plateau at the junction of the eastern part of the Inner Mongolia Autonomous Region with Shaanxi Province and the Ningxia Huizu Autonomous Region. The total area of the current Mu Us Sandy Land is about 32,100 km$^2$, accounting for 4.7% of the total desert area in China. The average annual precipitation from 1955 to 2021 was 271.1 mm, 63% of which was concentrated from July to September, resulting in heavy rains and storms in the summer (Ran et al. 2025). The annual average evaporation is 2226.8 mm, which is 8.2 times the rainfall, so droughts occur frequently. The average annual temperature over 66 years was 7.2 °C, with the lowest temperature being −24.6 °C on February 8, 1957, and the highest temperature being 30.6 °C on July 20, 2000. The Mu Us Sandy Land has a cold and dry winter and a humid and warm summer. Moreover, the average annual temperature from 1982 to 2014 has been consistently rising (as shown later in Fig. 3.4a) (Xiu et al. 2018).

The Mu Us Sandy Land is located around the northern limit of traditional rainfed agriculture, the 400 mm annual precipitation line, so both field crop production and livestock farming are practiced, and 13 types of landscapes are recognized based on the physico-geographical conditions and land use status (Wu and Ci 2002) (Fig. 3.1)—(1) grassland, (2) fixed sand dunes (Fig. 3.2), (3) semi-fixed sand dunes, (4) shifting sand dunes (Fig. 3.3), (5) complex of sand dunes and salinized and/or damp wetlands, (6) lower wetlands salinized and/or damp, (7) farmland, (8) abandoned farmland, (9) water body, (10) residential area, (11) shrub forest, (12) complex of sand dunes and willow lowland forests, and (13) artificial forest. Each ecosystem is affected by climate change and human activities, and large-scale desertification is progressing.



K. Yoshikawa et al., *Dryland Vegetation in the Sand-Shifting Environment*,
https://doi.org/10.1007/978-981-95-4735-7_3

**Fig. 3.1** Aerial view of the Mu Us Sandy Land taken from a kite

The Mu Us Sandy Land is located on the northwestern margin of China's monsoon zone, and its climate and environmental changes are closely related to the Southeast Asian monsoon that protrudes to the northwest and then retreats to the southeast, and its medium- to long-term fluctuations determine the pattern of sand movement in the Mu Us Sandy Land. That is, when the East Asian summer monsoon strengthens, the precipitation increases and the climate becomes warm and humid, and when the winter wind strengthens, the precipitation decreases and the climate becomes cold and dry. However, precipitation and temperature are not always linked, and not only warm and humid, cold and dry conditions appear, but also warm and dry, cold and humid conditions can occur.

For example, from 1990 to 2010, the temperature of the Mu Us Sandy Land showed a gentle upward trend, but the precipitation decreased (Fig. 3.4a), leading to increased surface evaporation and reduced effective soil moisture (Han et al. 2020). In particular, a severe drought struck from 1990 to 2000 (Fig. 3.4b), suppressing plant growth and advancing desertification. However, between 2010 and 2017, both temperature and precipitation increased year by year, resulting in a warm and humid

**Fig. 3.2** Fixed sand dunes

climate, which promoted plant growth, increased vegetation area, and led to the recovery of desertified land.

## 3.2 Origin and History of the Mu Us Sandy Land

The aeolian sandlands of the Ordos Plateau are mainly weathered bedrock from the Jurassic (201–145 million years ago) and Cretaceous (145–66 million years ago) periods, with wind-transported lacustrine deposits accumulating in depressions formed by the weathering of the bedrock (Wang et al. 2011). The Mu Us Sandy Land began to form during the Quaternary period (from the appearance of humans 2.59 million years ago to the present), and its transformation process is not a unilateral sand deposition, but a repeated cycle of the formation, fixation, and reduction of shifting sand dunes.

The formation of deserts and the recovery of desertified lands in arid and semi-arid regions have been repeated many times throughout the Quaternary period, but these phases could not be associated primarily with human impacts, because of the low population in that period. In the Ulan Buh Desert and Hobq Deserts, aeolian sand dune sequences have been deposited since about 580,000 years ago, and in the past 150,000 years, at least 27 alternating cycles of deposition of aeolian dune sands

**Fig. 3.3** Shifting sand dunes

and of fluvio-lacustrine facies or paleosols have occurred (Wang et al. 2008). The Mu Us Sandy Land was almost completed during the last glacial period (70,000–10,000 years ago) of the Pleistocene (the first half of Quaternary period), and during the climatic optimum of the Holocene (7000–5000 years ago), temperatures rose, leading to the development of moist vegetation; shifting sand was stabilized over the entire or most of the sandland, resulting in the appearance of fixed and semi-fixed sand dunes. Subsequently, during the human historical period, desertification due to aeolian sand began. In China, it has been confirmed by historical records and archaeological evidence that desertification has occurred many times over the past several thousand years.

The grasslands in the southern part of the Ordos Plateau were suitable for grazing, so nomads have lived there since ancient times. In 407 AD, the Huns founded the Daxia regime (407–431), and in addition to their nomadic life, they began farming and cultivating the grasslands. During the subsequent Tang Dynasty (618–907), a large number of troops were stationed in the Ordos region for frontier defense, leading to deforestation and burning of grasslands for cultivation, which severely damaged the ecological environment of the grasslands. Underneath the surface of the Ordos Plateau is a very thick layer of silt, and when the topsoil is destroyed and the silt is exposed, it is blown away by the wind, forming shifting sand dunes. During the Song Dynasty (960–1279), a “sand sea” stretching 350 km from Shanxi Province to the Ningxia Huizu Autonomous Region was formed. As a

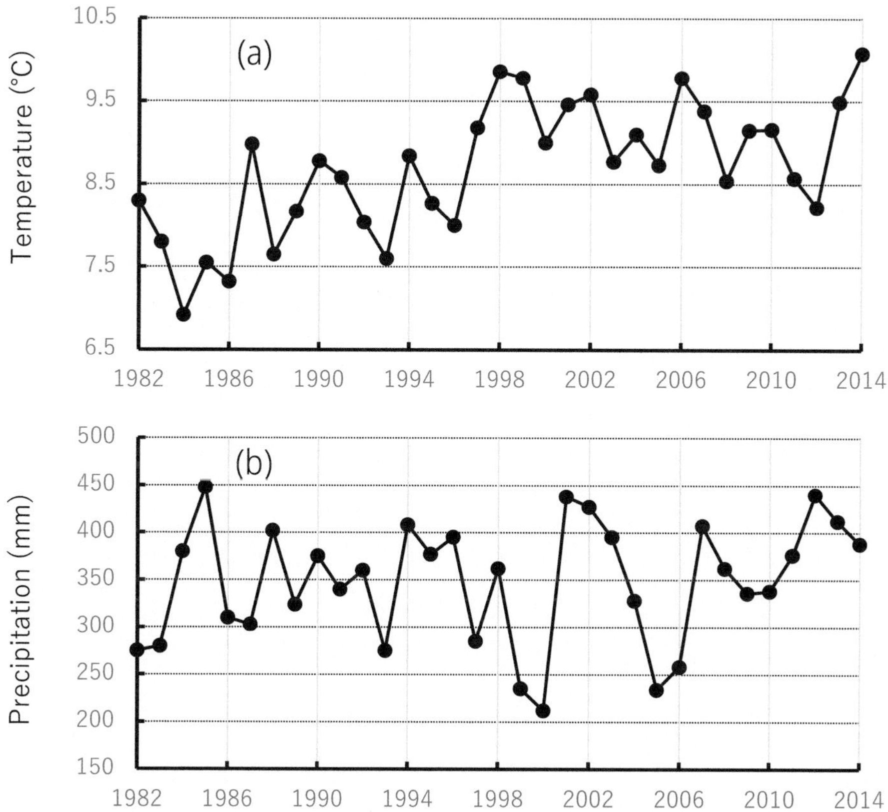


**Fig. 3.4** Changes in the annual mean temperature (**a**) and precipitation (**b**) in the Mu Us Sandy Land from 1982 to 2014. (Redrawn from Xiu et al. 2018)

countermeasure, the Great Wall, stretching over 7300 km long from the Yalu River in the east to Jiayuguan in the west, was built between 1465 and 1487 during the Ming Dynasty (1368–1644) (Fig. 3.5). However, the construction process significantly destroyed the ecological environment around the Great Wall, leading to desertification due to aeolian sand spread in the devastated grasslands and steppes. After the construction of the Great Wall, local governments encouraged farmers to cultivate the grasslands, and according to records, the farmland cultivated by farmers in the Yansui region reached about 600 ha. The military also cultivated nearby grasslands, but after several years of planting, the underlying sand was exposed to wind erosion, farmlands were abandoned, and even parts of city walls were buried by shifting sands (Deng et al. 2007).

Generally speaking, though the intensity of human activity has been increasing in this area since the Ming Dynasty, the Mu Us Sandy Land has not significantly expanded toward the southeast or southwest. The southern boundary of the Mu Us Sandy Land was controlled primarily by zonal natural factors such as rainfall,

**Fig. 3.5** Great Wall remaining on the outskirts of Yinchuan (Xinjian Uygur Autonomous Region), constructed during the Ming dynasty

temperature, vegetation, and soils. However, it is said that the 100 years before 1949, shifting dunes moved more than 40 km southward. In Jingbian County, Yulin City, 650 $km^2$ of farmland was buried in sand, about 270 $km^2$ of grassland suffered from aeolian desertification or salinization, 82 villages were encroached upon by shifting sands, and residents had to relocate to other lands. Yulin City, which originally earned its name from the elm forest, was forced to move three times due to the onslaught of moving sand.

In particular, records of desertification began to increase from the 1950s, and the rate of desertification was an average of 1560 $km^2$/year until the mid-1970s, but then desertification progressed rapidly, reaching 2100 $km^2$/year until the late 1980s (Wang et al. 2008). As a result, over the 35 years from 1958 to 1993, landscapes have changed significantly in the Mu Us Sandy Land. The proportion of shifting sand dunes increased from 30.2% to 44.5%, and semi-fixed sand dunes also increased from 10.9% to 21.4%. Meanwhile, fixed sand dunes decreased from 22.3% to 7.2%. These survey results were subsequently and extensively used in policy-making by the Chinese government as evidence supporting the hypothesis that desertification is caused by human activities.

The approximately 30 years from 1990 to 2017 can be divided into three periods. That is, the year 2000 was a turning point of the dynamic change of the land with different degrees of desertification. Before that, the desertified land area of the Mu

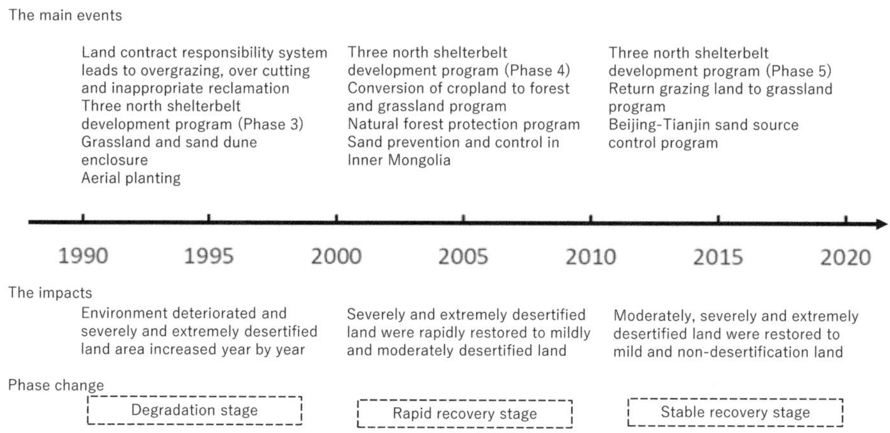


**Fig. 3.6** Timeline of human activities in the Mu Us Sandy Land. (Redrawn from Han et al. 2020)

Us Sandy Land was in the development stage and expanding spatially, but a rapid reversal recovery stage of desertification occurred from 2000 to 2010, followed by a slower stable recovery phase since 2010 (Han et al. 2020) (Fig. 3.6). As a result, the main types of desertification in 1990 were moderate (42.6%) and severe (35.4%), but by 2017, they had changed to moderate (45.0%) and light (32.0%), significantly reducing the overall degree of desertification. Of course, serious local deterioration and desertification was still occurring. But overall, desertification is recovering nationwide since 2000, and the decreasing trend has accelerated over the past 5 years. As a result, the total area of desertified land nationwide shrank by 1.45% over the 5 years from 2014 (Zan et al. 2023).

## References


Deng H et al (2007) Distribution of sand dunes and sand shifts along the southern fringe of the Mu Us desert since the Ming dynasty. Chin Sci Bull 52(22):3128–3138

Han et al (2020) Spatiotemporal dynamic evolution and driving factors of desertification in the Mu Us Sandy Land in 30 years. Sci Rep 10:21734

Ran B et al (2025) Analysis of soil moisture dynamics and its response to rainfall in the Mu Us Sandy Land. Water 17. https://doi.org/10.3390/w17010105

Wang X et al (2008) Desertification in China: an assessment. Earth Sci Rev 88:188–206

Wang T et al (2011) Deserts and Aeolian desertification in China. Elsevier

Wu B, Ci LJ (2002) Landscape change and desertification development in the Mu Us Sandland, Northern China. J Arid Envir 50:429–444

Xiu L et al (2018) Monitoring the response of vegetation dynamics to ecological engineering in the Mu Us Sandy Land of China from 1982 to 2014. Environ Monit Assess 190. https://doi.org/10.1007/s10661-018-6931-9

Zan GS et al (2023) Key data results and trend analysis of the sixth national survey on desertification and sandification. Forest Resources Management. https://doi.org/10.13466/j.cnki.lyzygl.2023.01.00

# Part II
# The Environment of the Mu Us Sandy Land

## Introduction

In the first half of this part (Chap. 4), we will explain the climate, geology, topography, and hydrological environment of the Mu Us Sandy Land. We will focus on the origins, evolution, and current state of the shifting sand dune environment. For this purpose, we will introduce long-term (1990–2016) observed meteorological data. In the hydrological environment, the formation of dry sand layers and the evapotranspiration of vegetated area are emphasized as characteristic issues affecting groundwater level fluctuations.

In the next chapter (Chap. 5), we will explain the ecological characteristics of the plants (woody and herbaceous) of the Mu Us Sandy Land. In particular, we will focus on the impact of the shifting sand environment on plants. We will also explain the ecological and physiological characteristics of plants in shifting sand dunes, fixed sand dunes, and interdune lowlands comparatively.

# Chapter 4
# Physical Environment

**Nobuhito Ohte**

## 4.1 Location, Geographical Features, and Land Use

The Mu Us Sandy Land is spread over the northern part of the Ordos Plateau, located in the middle reaches of the Yellow River. It spans a range of 38–39° north latitude, 107–110° east longitude, and an altitude of 1000–1300 m. Its area is about 35,000km$^2$. The annual precipitation in Wushenzhao, located in the central part of the Mu Us Sandy Land, is 324 mm, and the average annual temperature is 7.2 °C (Fig. 4.1). According to Köppen's climate classification, the annual precipitation falls below the dry limit, and the average annual temperature falls below 18 °C, so it is classified as a cold steppe climate, and vegetation such as shrubs and herbs adapted to semi-arid environments is established.

The strong northwesterly seasonal winds that blow from March to June cause intense sand flying. The formation of a significant area of shifting sand is largely due to these seasonal winds. The geological origin of this shifting sand was described to be the sandstone layer formed in the Mesozoic era and the thick sedimentary layer formed in the Cenozoic era, and the formation of the shifting sand area is believed to have progressed in the Quaternary period when the climate became drier (Lanzhou Desert Research Institute of the Chinese Academy of Sciences 1974; Li and Chen 1983; Zhu et al. 1981; Chow et al. 1988).

By the way, the area this book focuses on is called the Mu Us "Sandy Land." In northern China, the main landscapes in arid and semi-arid areas are lands including sandy lands. Among them, areas with strong dry conditions and where shifting sand dunes are the main landscape are called deserts, and areas that are more humid, where steppe vegetation covers the sandy land surface, and where fixed sand dunes or semi-fixed sand dunes occupy a certain area are called sandy lands. In other words, in "sandy lands," quicksand area is still a major landscape, but it implies that various vegetation areas are included. For various socially used lands such as pastures and cultivated lands covered by vegetation, greening and afforestation aimed at stabilizing the quicksand mentioned later are applied to the surroundings (refer to Part IV).



K. Yoshikawa et al., *Dryland Vegetation in the Sand-Shifting Environment*,
https://doi.org/10.1007/978-981-95-4735-7_4

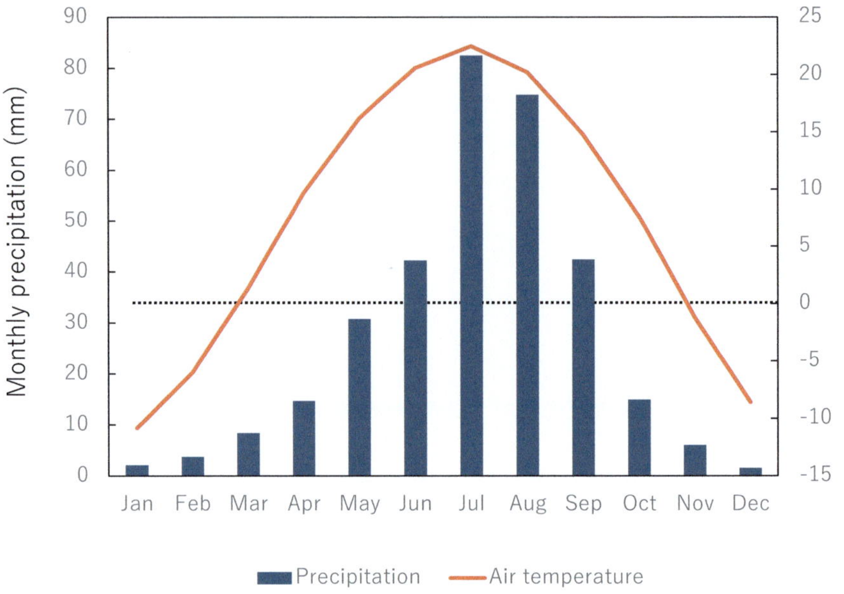


**Fig. 4.1** Monthly mean are temperature and monthly precipitation. Meteorological observations were conducted at the Wushenzhao Meteorological Observatory (WMO station No. 53547). Monthly values are averages from 1996 to 2017 (ref.). Some missing data are included for some years

In the range called the Mu Us Sandy Land, there are several references that describe how the area of quicksand and vegetated land is proportioned, and those temporally changed. Ma et al. (1998) and Zheng (2009), with data before the 1990s, state that the proportions were as follows: shifting sand dunes: 31.6%, semi-fixed sand dunes: 36.5%, fixed sand dunes: 31.9%. As shown in the introduction, by the early 1990s, the proportions were as follows: shifting sand dunes: 44.5%, semi-fixed sand dunes: 21.4%, fixed sand dunes: 7.22%; the shfting sand dunes had expanded (Wu and Ci 2002). They state that this expansion was the result of desertification due to the conversion to farmland and pastureland that has continued since the 1950s. As for the situation in the 2000s, Liang and Yang (2016) state that the proportions were as follows: moving sand dunes: 11.68%, semi-fixed sand dunes: 18.04%, shrubs/farmland: 30.91%, pastureland/grassland: 38.94%. Since the early 2000s, moving sand dunes have significantly decreased and vegetated areas have increased. This trend since the 2000s has been shown in several recent analyses using remote sensing data (Li et al. 2022; Shen et al. 2022). Liang and Yang (2016) state that vegetation coverage has significantly increased in the eastern part due to ecosystem restoration activities, but on the other hand, since 2000, a "new nominal farmland development" for the purpose of restoring the ecosystem has begun, and there are reports that a transition from traditional rain-fed agriculture to irrigation agriculture dependent on groundwater has occurred (Yong-Hai et al. 2017). This suggests that the decrease in moving sand dunes since the 2000s was substantially due to the element of conversion to cultivated land.

## 4.2 Geophysical Features

### *4.2.1 Geology and Soil*

The Ordos Plateau, where the Mu Us Sandy Land is located in the center, has an old foundation, believed to be established in the Precambrian era (1.7 billion years ago). Therefore, the terrain is flat and forms a plateau. In the Triassic period of the Mesozoic era, the surroundings of the Ordos became an inland basin, and a sandstone layer that became the bedrock was formed due to the inflow of sand and gravel from the surroundings during the Jurassic and Cretaceous periods that followed. From the end of the Jurassic period, uplift and folding occurred in the surroundings due to the influence of the Yanshan-Himalayan orogeny that began, but in the tertiary of the Cenozoic era, the Ordos was uplifted, and the terrain changed from a depositional terrain to the current plateau-like terrain. Even during the period until the uplift ended, sedimentation of soil continued, and a thick quaternary layer was formed on the sandstone, which is the bedrock (Ren et al. 1986).

The current Mu Us Sandy Land is roughly divided into the southeastern part covered by a thick quaternary deposit layer and the central and western part where the bedrock layer formed in the Jurassic and Cretaceous periods of the Mesozoic era is exposed. It has been shown that the quaternary deposit layer includes not only water-formed deposit layers but also wind-formed deposit layers (Dong et al. 1983), indicating that there were periods of advanced dryness. In particular, it is believed that the quicksand area was most widely spread during the dry period in the first half of the quaternary, and most of the origin of the sand is quaternary deposits.

From the above-described geological conditions, the soil in most of the Mu Us Sandy Land is sandy, and soil containing some organic matter and fine soil materials develops in parts covered by stable vegetation.

### *4.2.2 Shifting Sand and Landscapes*

As mentioned above, the sandstone layer in the area covered by the thick quaternary deposit layer in the southeastern part is considered to be the main source of the current moving sand (Ren et al. 1986), and the source of the moving sand in the high plateau area in the central and western part is the sandstone deposited and formed in the Mesozoic era. In any case, these sands are not brought from areas other than the Ordos Plateau, but are believed to be generated by the weathering of the deposit layer in this area.

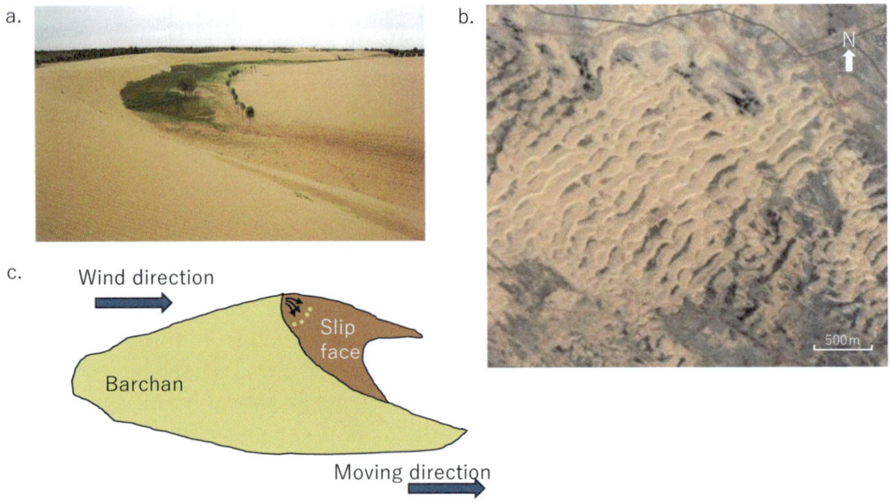


**Fig. 4.2** (**a**) Barchan formed by moving sand (Photographed by Yoshikawa). (**b**) Barchans formed by northwest winds (Satellite image from Google Earth 2025). (**c**) Conceptual diagram of the barchan formation

Quick sands often form crescent-shaped dunes in a continuous manner, much like in photographs (Fig. 4.2). This type of dune is called a "barchan," which protrudes two corners on the leeward side, and the angle of the leeward slope (slip face) becomes the angle of repose. The sand on the windward side gathers due to the wind, and the deposition angle is around 15°. Generally, the formation of barchans in sand deserts is actually very rare, and it requires conditions such as uniform grain size of quicksand, relatively limited supply, and wind blowing in a certain direction over a certain long period of time. The aforementioned strong seasonal winds from winter to spring in the Mu Us Sandy Land play a significant role in this geomorphological formation. The size of the dunes varies, but it is not uncommon for the relative height from the base of the dune to exceed 10 m. The speed of the moving dunes is about 1 m/year, and they are moving from northwest to southeast while maintaining their shape (Kobashi and Okumura 1989, Fig. 4.3).

The depth of the groundwater surface at the base of the dune is often 0.5–1 m and at most a few meters. In other words, the soil environment of the interdune lowlands between the dunes is relatively moist, and herbs and shrubs can grow, and it is used as pasture (Kobashi and Kobayashi 1989). This landscape, where dunes and interdune lowlands appear alternately, is common in the southeastern part where there is a lot of sand derived from quaternary deposits, and it forms the typical landscape of this region.

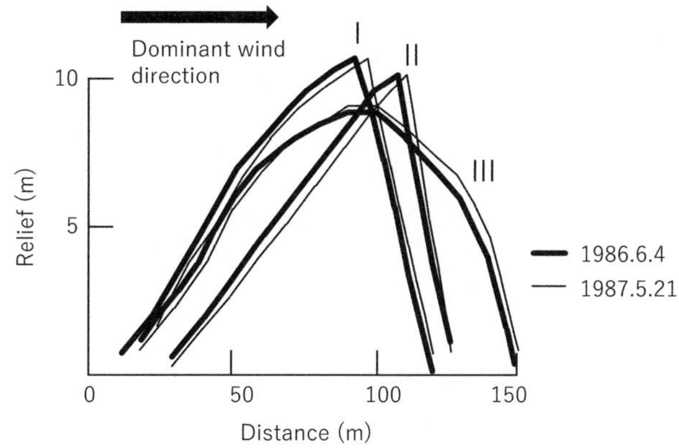


**Fig. 4.3** Annual shift of sand dunes. I–III are the different sample dunes. (Kobashi and Okumura 1989)

## 4.3 Climate and Hydrology

### *4.3.1 Precipitation and Groundwater Dynamics*

In the cold steppe climate of the Mu Us Sandy Land, the seasonal variation in precipitation is clear, showing a typical Asian monsoon type of seasonal change with more in the summer. Seventy eight percent of the annual precipitation is concentrated from June to September, and the winter precipitation is small. The monthly average temperature is below freezing for the 4 months from November to February, and soil freezing can be seen during this time (Fig. 4.1).

Looking at the long-term changes in precipitation, the minimum value from 1965 to 2001 is 150 mm, the maximum value is 530 mm, and there is a difference of nearly 90% of the average value. This change in precipitation suggests that it leaves a history of changes in the region's water storage (mostly groundwater storage) over the years (Fig. 4.4).

As mentioned above, in typical dune areas, the groundwater surface is often at a height of 0.5–1 m from the base of the dune. Therefore, during the summer when precipitation is concentrated, the soil of the interdune lowlands remains moist, and ponding may occur on the ground surface. The fluctuation of the groundwater level in these relatively moist interdune lowlands is strongly affected by the evapotranspiration of the growing interdune lowlands.

Figure 4.5 shows the fluctuation of the groundwater level in the interdune lowlands where vegetation is abundant in growing season. The fluctuation in winter (December) is not clear, but the decrease due to the effect of vegetation evapotranspiration begins in spring (April) and continues until the beginning of winter in October. During the rainy season (June to September), there are several large rainfall events that cause an overall rise in the groundwater level (Fig. 4.5).

The groundwater level on sunny days from April to October, when there is vegetation evapotranspiration, decreases during the day and rises at night, including

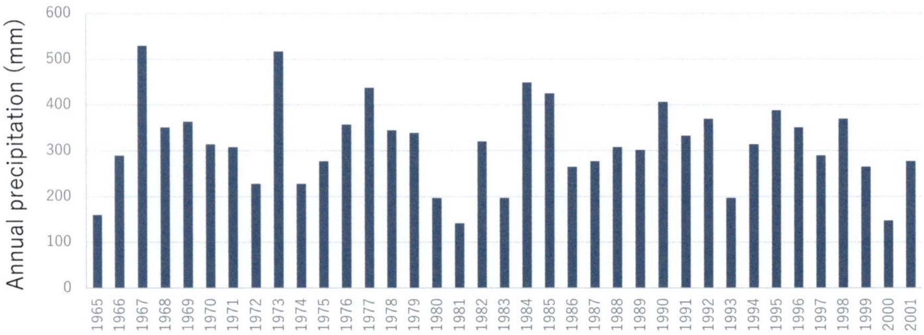


**Fig. 4.4** Long-term change of the annual precipitation in the Mu Us Sandy Land. The source data was provided from the Wushenzhao Meteorological Observatory (WMO station No. 53547)

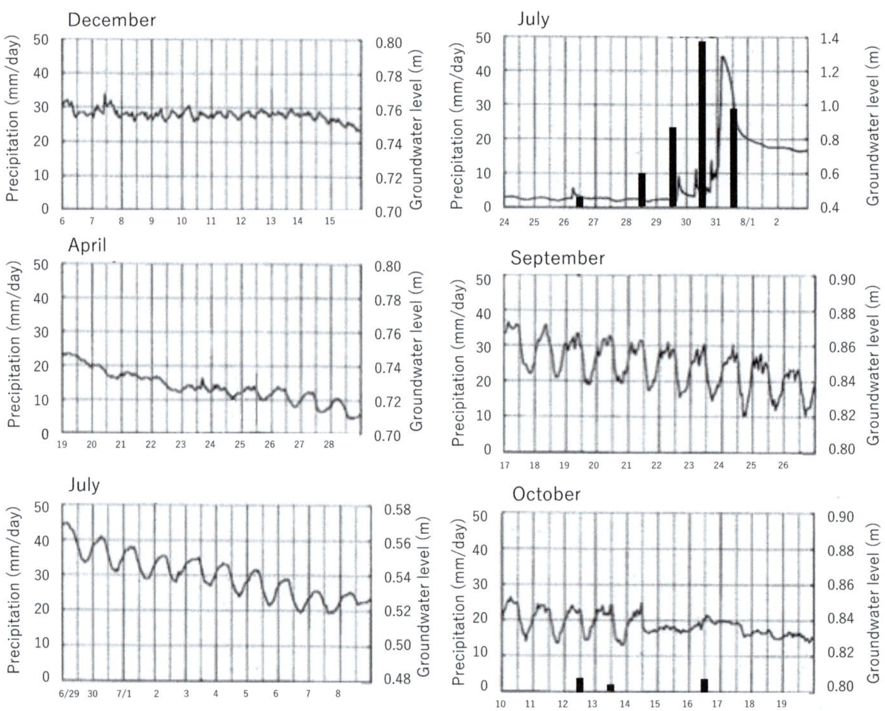


**Fig. 4.5** Seasonal and diurnal variation in the precipitation and groundwater level (from the bottom of the observation well) at the vegetated part of the interdune lowland

daily fluctuations, and this can be interpreted as soil moisture consumption due to vegetation transpiration causing groundwater level decline through capillary rise. The rise in water level at night cannot fully compensate for the decrease due to evapotranspiration, and the groundwater level will be lower than the same time the

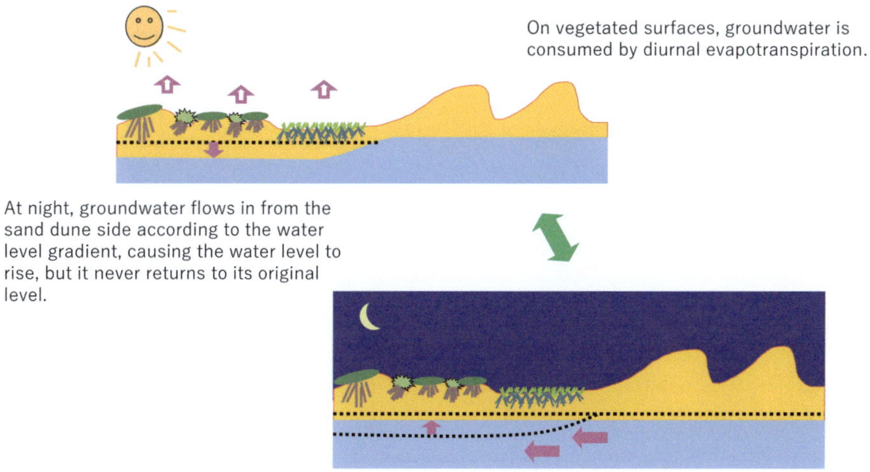


**Fig. 4.6** Mechanisms of groundwater level fluctuations in the vegetated part

previous day. The rise in groundwater level at night can be considered as inflow according to the hydraulic gradient from the surrounding dune land where the groundwater level has not dropped (Fig. 4.6).

This evapotranspiration and groundwater dynamics in the interdune lowlands are supported by a survey of stable isotope ratios of groundwater (Kabeya et al. 2002). Figure 4.7 shows the results of measuring the electric conductivity (EC), salt ($Na^+$, $Cl^-$) concentration, and stable isotope ratio ($\delta^{18}O$, $\delta^2H$) of groundwater in a transect including dunes and vegetation (herbs) in the interdune lowlands. The stable isotope ratio of precipitation globally has the following relationship found by Craig (1961):

$$\delta^2\mathrm{H} = 8 \bullet \delta^{18}\mathrm{O} + 10 \tag{4.1}$$

The X-axis shows $\delta^{18}O$ and Y-axis shows $\delta^2H$; the straight line indicated by the above equation is called the Meteoric Water Line. Dansgaard (1964) pointed out that when water undergoes strong non-equilibrium evaporation, the isotopic ratio of the remaining water deviates from the above equation, and he expressed this effect as the *d* value (or deuterium excess) in the equation below:

$$d = \delta^2\mathrm{H} - 8 \bullet \delta^{18}\mathrm{O} \tag{4.2}$$

The *d* value corresponds to the Y-intercept of a line with a slope of 8 passing through the observed values on the above diagram, and the intercept of the MWL is 10. When subjected to strong evaporation as above, the *d* value shifts to a value less than 10. In Fig. 4.7, both the inorganic salt concentration and the stable isotopic ratio of water concentrate when subjected to evaporation, and both are high in groundwater in interdunal lowlands, indicating that the groundwater was affected by evaporation.

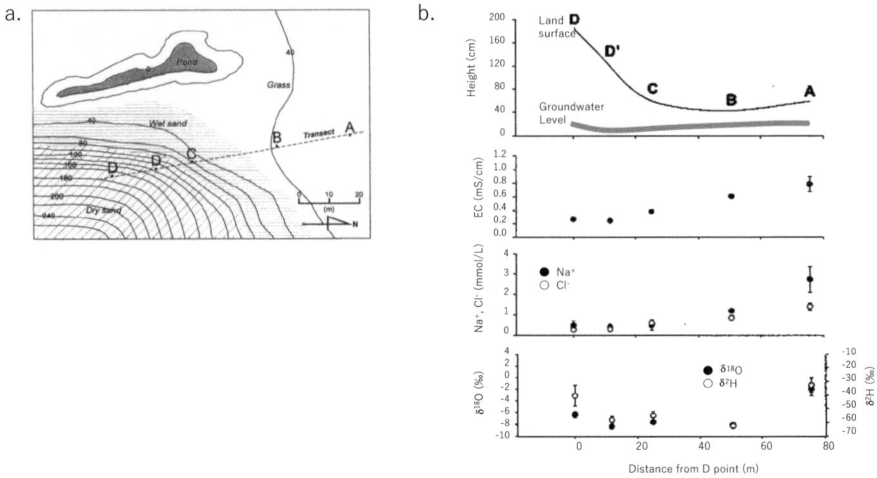


**Fig. 4.7** (**a**) Ground surface condition and topography of the observation transect. (**b**) Electric conductivity (EC), $Na^+$ and $Cl^-$ concentration, and $d^2H$ and $d^{18}O$ values of the groundwater along the transect. (Kabeya et al. 2002)

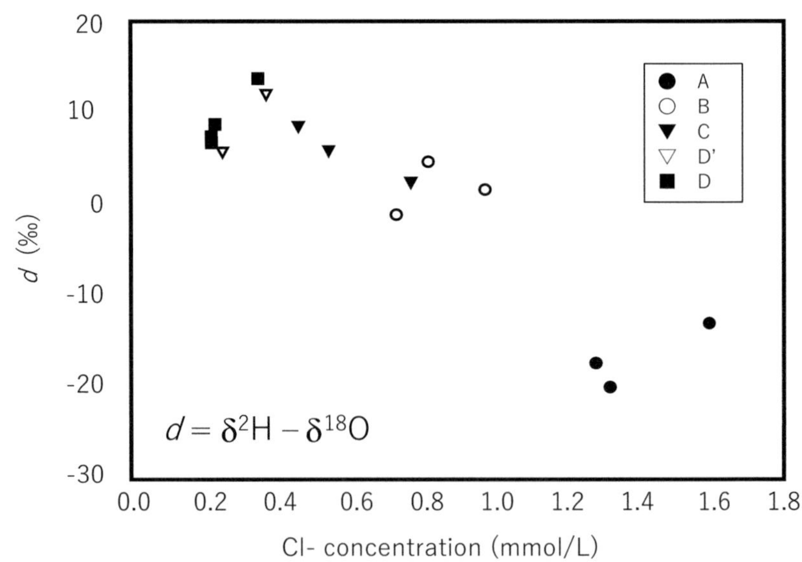


**Fig. 4.8** Relationship between $Cl^-$ concentration and *d* (deuterium excess) value. A—D and D′ indicate the sampling points of groundwater shown in Fig. 4.7

Also, Fig. 4.8 shows the relationship between $Cl^-$ concentration and *d* value, and it is clear that both are changing in tandem. The fact that plant transpiration does not affect the *d* value suggests that groundwater consumption in interdunal lowlands due to evapotranspiration depends more on evaporation than on transpiration.

In typical dune areas, the groundwater level is often 0.5–1 m relative to the base of the dunes. During the summer, precipitation is concentrated, so the soil in the interdunal lowlands is moist at this time, and waterlogging may occur on the surface. This part is the main evaporation surface of the area, and it is clear from the above that evaporation from the dune itself is small compared to this.

### *4.3.2 Evaporation and Dry Sand Layer*

It is clear from observations of radiation balance and heat balance on the surface that the amount of evaporation from the dune itself is not large. Figure 4.9 shows the daily changes in (a) radiation (solar radiation and net radiation) and (b) heat balance on the surface covered by several types of vegetation (c) moist surface with low herbs, (d) cover by *J. sabina*, (e) meadow (*Corethrodendron lignosum* (=*Hedysarum mongdicum*)) around the dune (Nakamura and Kobashi 1991). The net radiation is the balance of radiation coming in from the sun and the sky and radiation going out from the ground. Shortwave radiation from the sun is partially reflected on the ground, and longwave radiation is radiated from the heated ground to the sky, but some is reflected by clouds and returns to the ground. The net radiation obtained from these balances, its energy is consumed by evaporation of water, heating of the atmosphere, and heating of the soil on the ground. The significantly lower net radiation in dune areas compared to solar radiation is due to the extremely high reflectivity of the dry, white sandy surface.

As a result, the net radiation (net energy) brought to the surface is smaller than the vegetation area, and almost all of it is consumed to heat the atmosphere (Fig. 4.9). In other words, hardly any evaporation occurs from the dune. On the other hand, the heat energy consumed varies depending on the type of vegetation cover, and on the moist surface with low herbs, most of the energy is used for water evaporation, and where *J. sabina* covers, the energy is used to heat the atmosphere. In the pasture, the heat consumption was intermediate, with about half being consumed by evapotranspiration. *J. sabina* is a representative vegetation in fixed dunes, a long-lived woody plant, and its nature to suppress evapotranspiration is reflected in the observation results. The remaining herbaceous vegetation showed significantly more evapotranspiration than the dunes. From the above, it is shown from observations that in areas with moving dunes, large evapotranspiration occurs mainly in interdunal lowlands, and the amount of evaporation on the dune surface is small. This is consistent with the interpretation of groundwater dynamics in the previous section.

Then how can we explain the mechanism of low evaporation from sand dunes? On the surface of sand dunes, a very dry layer of sand is formed when there is a

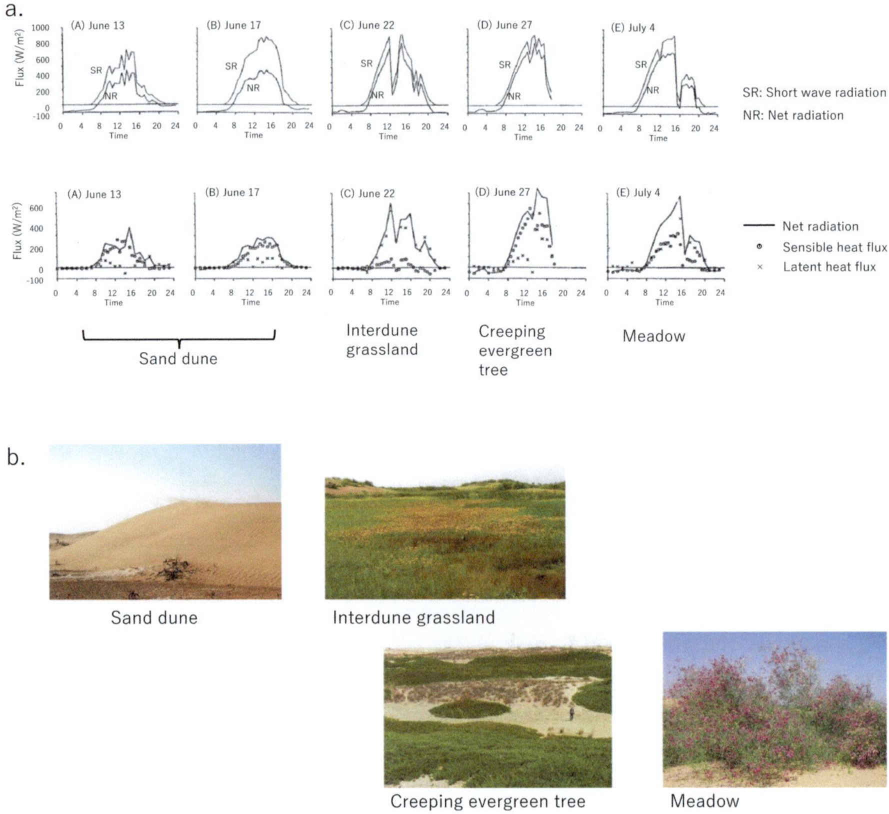


**Fig. 4.9** (**a**) Short wave radiation, net radiation, sensible heat flux, and latent heat flux at various land surfaces. The observation was carried out in June and July 1991. (A) and (B) were observed at sand dunes. Interdune grassland (C) consisted of multiple grass species. Creeping evergreen tree (D) was *Juniperus sabina* (see Chap. 9). Meadow (E) consisted of *Corethrodendron lignosum* (=*Hedysarum mongdicum*) plantation. (**b**) Surface condition of each landscape. (Photographed byK. Yoshikawa and N. Yamanaka)

period without rainfall (Fig. 4.10). Inside the dry sand layer, water only exists in the form of water vapor, and water movement only occurs in the gas phase. The dry sand layer grows from the surface downward, but it hinders the diffusion and movement of water evaporated from the bottom of the layer, so the amount of water vapor released from the surface to the atmosphere decreases as the layer becomes thicker. As a result, when it reaches a certain thickness, the amount of evaporation from the surface becomes zero. Rainwater that falls when the dry sand layer is formed is first consumed to eliminate the dry sand layer, and then more rainwater vertically infiltrates downward. This phenomenon has been modeled by Ohte et al. (1989) and He and Kobayashi (1998), and estimates have been made of the change in evaporation from the sandy surface and the amount of rainwater infiltrating below the dry sand layer (Ohte et al. 1989).

**Fig. 4.10** Dry sand layer. (Photographed by Yoshikawa)

Figure 4.11 shows the schematic diagram of the simulation model of the dry sand layer by Ohte et al. (1989). Assuming no evaporation occurs during rainfall, the equilibrium between the thickness of the dry sand layer, *D* and the evaporation rate from the sand surface, *Es* or rainfall rate, *R* are written as:

$$Es \text{ or } R = (\theta_r - \theta_0) \bullet \partial D / \partial t \tag{4.3}$$

Where $\theta_r$ is the water content on which liquid water transfer stops, and $\theta_0$ is the water content of the dry sand layer. Evaporation rate from the sand surface is restricted by the dry sand layer, because the dry sand layer prevents liquid water from moving to the ground surface. The relationship between evaporation and dry sand layer thickness is as follows:

$$Es = Ep \bullet \exp(-\alpha \bullet D) \tag{4.4}$$

Where a is $\alpha$ constant. The water flux *Re* supplied to layers deeper than the dry sand layer is expressed as follows.

$$Re = R - D \bullet (\theta_r - \theta_0) \tag{4.5}$$

Observations and model simulations show that evaporation, which starts from the sandy surface being sufficiently wetted by rainfall, becomes close to zero within a few days from the start, and during this time, the dry sand layer reaches a thickness of about 7–8 cm.

Table 4.1 shows the estimated evaporation from the sandy surface and the amount of rainwater infiltration below the dry sand layer under certain precipitation conditions for 2 years. In these 2 years, the proportion of evaporation to precipitation in a dry year was 31.5%, and that in a wet year was 19.0%. This difference in evaporation was found to depend not on the amount of precipitation, but on the number of

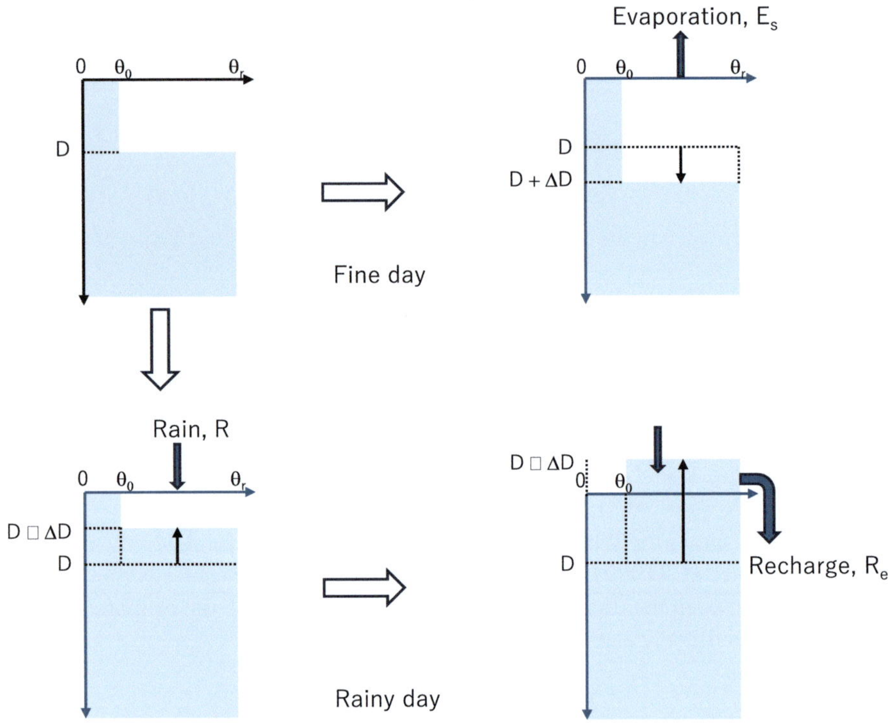


**Fig. 4.11** Schematic diagram of the simulation model for sand surface evaporation and dry sand layer thickness. *R* Rainfall, $E_s$ Snad surface evaporation, $R_e$ Water charge into infiltration zone, $q_r$ Minimum water content when the liquid water movement can occur, $q_0$ Water content of dry sand layer, *D* Thickness of dry sand layer

**Table 4.1** Calculated water budget on sand dune surface using the simulation model for sand surface evaporation and dry sand layer thickness

| Period | Observed precipitation (mm) | Observed potential evaporation (mm) | Calculated evaporation from soil surface (mm) | Calculated water charge under the dry sand layer (mm) | Number of raining days (days) |
|---|---|---|---|---|---|
| July 1, 1986–June 30, 1987 | 183.1 | 1644.4 | 57.7 | 129.1 | 52 |
| July 1, 1987–June 30, 1988 | 269.4 | 1766.3 | 51.1 | 219.5 | 40 |

rainfall days. Regardless of the amount of precipitation, the infiltration amount reaches 70–80% of the precipitation, and it is found that the sandy surface plays a large role in groundwater recharge. This contrasts with the active evapotranspiration in the inter-hill lowlands, and overall, it is inferred that groundwater is recharged in the sand dunes, and evapotranspiration in the vegetated inter-hill lowlands returns most of it to the atmosphere.

### *4.3.3 Estimation of Regional Water Budget*

Based on the information about the characteristics of the evaporation process of sandy and vegetated areas as described above, it is possible to estimate the water balance of the entire Mu Us sandy area. The general formula for water balance can be written as follows:

$$\begin{aligned}[\text{Precipitation}] &= [\text{Runoff}] + [\text{Evapotranspiration}] \\ &+ [\text{Change in soil and groundwater storage}]\end{aligned} \tag{4.6}$$

If the calculation of the water balance is done on an annual basis under average precipitation conditions, it is generally possible to approximate [Change in soil water and groundwater storage] to zero. Also, considering that there are no rivers with constant runoff in the area, [Runoff] can also be approximated to zero. In other words, a rough estimate would predict that [Annual precipitation] is nearly equal to [Annual evapotranspiration].

According to the observations by Nakamura and Kobashi (1991) shown in Fig. 4.9, the average evapotranspiration rate in typical vegetated areas in lowlands and fixed sand dunes is 5.3 mm/day at the peak of summer and 0.8 mm/day in sand dunes. The period when evapotranspiration is most pronounced is 6 months (183 days) from mid-April to mid-October according to Fig. 4.5, and considering the seasonal variation in which evapotranspiration is high in summer and slightly lower in spring and autumn, multiplying the daily evapotranspiration rate by the number of days and 0.8 for each of the above land surfaces gives 776 mm and 117 mm, respectively. If the area ratio of sand dunes to vegetated areas is 3:7, the total evapotranspiration rate is 315 mm. Since the annual average precipitation is 324 mm (in the former section), this estimate is considered to capture the overall water balance well.

## References

Chow C, Matsuda A, Kamichika M (1988) Overview of the Mu us desert development and research centre, China (1) Overview of the natural environment in the experimental site (in Japanese). Sakyu Kenkyu, 35(1):41–47

Craig H (1961) Isotopic variations in meteoric waters. Science 133(3465):1702–1703. https://doi.org/10.1126/science.133.3465.1702

Dansgaard W (1964) Stable isotopes in precipitation. Tellus 16(4):436–468. https://doi.org/10.3402/tellusa.v16i4.8993
Dong G, Li B, Gao S, Wu S, Shao A (1983) Quaternary wind-formed sand on the Ordos plateau (in Chinese). Geographica Sinica 38(4):341–347
He W, Kobayashi T (1998) A rational parameterization of evaporation from dry, bare soil. J Meteorol Soc Japan Ser II 76(6):955–963. https://doi.org/10.2151/jmsj1965.76.6_955
Kabeya N, Ohte N, Sugimoto A, Yoshikawa K, Wang L-h (2002) Effects of the ground cover conditions on horizontal profiles of chloride concentration, oxygen and hydrogen stable isotope ratios of groundwater in Mu Us Desert, China (in Japanese with English summary). J Japan Soc Hydrol Water Resour 15(1):13–22. https://doi.org/10.3178/jjshwr.15.13
Kobashi S, Kobayashi T (1989) Natural vegetation and revegetation technology. In: Elucidation of the mechanism and dynamics of desertification in China's arid regions – Basic research on desert greening and agricultural development in the Mu Usu Desert. Inner Mongolia Desert Development Research Group
Kobashi S, Okumura T (1989) Sand drift and control technology. In: Elucidation of the mechanism and dynamics of desertification in China's arid regions – Basic research on desert greening and agricultural development in the Mu Usu Desert
Li H, Chen D (1983) Sources of sands and desertification issues. In: P. U. The Department of Geography (ed) Soil geography and soil resources: natural conditions of the Mu Us Sandy area and their improved utilization (in Chinese). China Science Publishing & Media Ltd, pp 5–16
Li M, Zhong S, Luo Y, Liu Q, Li X (2022) A study of the change in surface parameters during the last four decades in the Mu Us Desert based on remote sensing data. Remote Sens 14(16)
Liang P, Yang X (2016) Landscape spatial patterns in the Maowusu (Mu Us) Sandy Land, northern China and their impact factors. Catena 145:321–333. https://doi.org/10.1016/j.catena.2016.06.023
Ma S, Ma Y, Yao H, Wang L, Yao Y (1998) Desertology. Inner Mongolia People's Publishing House
Nakamura A, Kobashi S (1991) Energy budget observation with the eddy correlation method in the Mu Us Shamo Desert. Studies on control of desertification and development of agriculture at arid land areas in China 1991. In: International symposium for Japan and China Joint Research, Tottori, Japan
Ohte N, Suzuki M, Kobashi S (1989) Influence of soil moisture conditions on the growth of Salix matsudana Koidz. In Mu Us Shamo desert, Inner Mongolia, China (I) (in Japanese with English summary). J Jpn Soc Reveget Technol 15(4):20–30. https://doi.org/10.7211/jjsrt.15.4_20
Ren J, Chunfat J, Zhengkun Z, Deok-yu C (1986) Geological structure of China and its developmental history (in Japanese translated by Y. Fujita and X. Long). Tsuiji Shokan
Research, L. I. o. C. a. D (1974) Introduction to Chinese deserts (in Chionese). China Science Publishing & Media Ltd
Shen B, Wei Y, Ma L, Xu D, Ding L, Hou L, Qin Q, Xin X (2022) Spatiotemporal changes and drivers of fractional vegetation cover in Inner Mongolia grassland of China (in Chinese with English summary). Trans Chin Soc Agri Eng 38(12):118–126. https://doi.org/10.11975/j.issn.1002-6819.2022.12.014
Wu B, Ci LJ (2002) Landscape change and desertification development in the Mu Us Sandland, Northern China. J Arid Environ 50(3):429–444. https://doi.org/10.1006/jare.2001.0847
Yong-hai, Hoshino B, Suriga, Sasamura T, Umegaki K, Nayinta (2017) The 'new agricultural land reclamation' current situation and topic regarding Inner Mongolia semi-farming semi-animal husbandry – a case study of village level in Chifeng City. J Arid Land Stud 27(1):9–16. https://doi.org/10.14976/jals.27.1_9
Zheng W (2009) Sandy desert and its control in China. Science Press
Zhu Z, Wu Z, Liu S, Di M (1981) Introduction to the Chinese Desert (in Chinese, Revised Edition). China Science Publishing & Media Ltd

# Chapter 5
# Biological Environment

**Norikazu Yamanaka**

## 5.1 Vegetation and Characteristic Land Morphology of the Mu Us Sandy Land

The Mu Us Sandy Land belongs to arid to semi-arid regions due to climatic conditions, and the vegetation is basically dominated by plants that are adapted to dry conditions. However, in the Mu Us Sandy Land, as the precipitation decreases from southeast to northwest, the vegetation changes accordingly. For this reason, the Mu Us Sandy Land can be divided into three regions with different vegetation. In the northwestern region, desert-steppe and steppe-desert vegetation can be seen, which are established under very dry conditions, and plants with strong resistance to drought predominate. In the central to southeastern part, typical steppe vegetation is seen that occupies most of the Mu Us Sandy Land. In the southeastern edge, vegetation with transitional characteristics between typical steppe and forest steppe develops. The vegetation of the typical steppe consists mainly of xerophytic and mesophytic plants, while in the northwest desert-steppe or steppe-desert, more drought-resistant xerophytic plants predominate. Typical plant species that make up the grassland vegetation (typical steppe) of the Mu Us Sandy Land are plants that belong to genus *Stipa* of Poaceae, such as *Stipa breviflora* and *Stipa bungeana*, which represent the temperate grasslands of Asia. In addition, many plants of the *Artemisia* of the Asteraceae family are seen, with the semi-shrub *Artemisia ordosica* being a representative example (Chen 1983; Kobayashi 1994; Wang et al. 2011b).

In the Mu Us Sandy Land, 401 species of higher plants from 68 families and 224 genera have been recorded. The Asteraceae family is the most abundant, with 57 species distributed throughout the region. The next most abundant is the Poaceae family with 50 species. The third is the Fabaceae family with 39 species (Chen 1983).

The landscapes that make up the Mu Us Sandy Land include (1) grassland, (2) fixed dunes, (3) semi-fixed dunes, (4) shifting dunes, (5) combination of dunes and saline wetlands, (6) saline low wetlands, (7) farmland, (8) abandoned farmland, (9)

 
K. Yoshikawa et al., *Dryland Vegetation in the Sand-Shifting Environment*,
https://doi.org/10.1007/978-981-95-4735-7_5

water surface, (10) residential area, (11) shrub forest, (12) lowland forest of dunes and willows, and (13) artificial forest (Wu and Ci 2002).

The most characteristic landscape classification in the Mu Us Sandy Land is the dune landscape, which includes fixed dunes, semi-fixed dunes, shifting dunes, and a combination of dunes and saline wetlands. Also, the Mu Us Sandy Land is a typical region in China where desertification due to wind erosion is progressing strongly (Wang et al. 2011b), and the movement of sand due to desertification has a significant impact on the plants that grow there. As a landscape element, in fixed dune areas where vegetation covers and prevents sand movement, if vegetation degradation occurs, sand movement becomes active, and fixed dune areas change to semi-fixed dune areas and then to shifting dune areas.

Here, we will focus on the movement of sand, a characteristic of the Mu Us Sandy Land, and explain the vegetation characteristic of each landscape, focusing on fixed dune areas, semi-fixed dune areas, shifting dune areas, and interdune lowlands existing between dunes, centered on Wushen Banner in the Inner Mongolia Autonomous Region (Fig. 5.1).

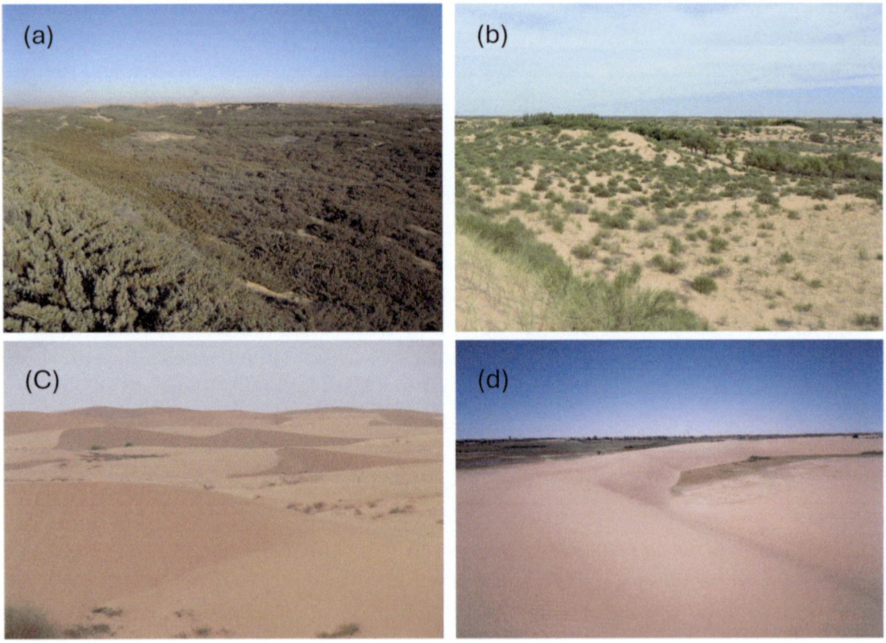


**Fig. 5.1** Typical dune landscapes in the Mu Us Sandy Land (**a**) Fixed dunes. (**b**) Semi- fixed dunes. (**c**) Shifting dunes. (**d**) Combination of dunes and saline wetlands. (Photos by K. Yoshikawa, Okayama Univ.: **a**, **b**)

## 5.2 Plants on Fixed Dunes

In fixed dune areas, the surface of the dunes is covered with vegetation, so even when the wind blows, there is almost no movement of sand. The surface is covered with vegetation, resulting in dunes with gentle undulations. As of 1993, fixed dune areas accounted for 7.22% of the total area of the Mu Us Sandy Land, but in 1958, it was 22.33%, indicating that the area is decreasing (Wu and Ci 2002).

In the fixed dune areas of the Mu Us Sandy Land, many plants that are resistant to drought and grow well in sandy areas are seen. There are few grasslands dominated by gramineous plants, but plants of the *Artemisia* genus of the Asteraceae family and shrubs of the Fabaceae family are widely distributed. In areas where the state of fixed dunes has been stable for a long time, communities of the shrub *Juniperus sabina*, belonging to the Cupressaceae family, develop, covering the surface of the dunes like a carpet (Fig. 5.2). In fixed dune areas, because vegetation develops, there is a lot of organic matter in the soil, and in the past, it was a target of development and its area has decreased, and the existing communities are protected. The *J. sabina* that represents the vegetation of the fixed dune area in the Mu Us Sandy Land is considered a climax species in vegetation succession and forms a stable community over a long period of time. The main inland distribution of this species in eastern Eurasia is in northern China, from Urumqi to Lanzhou and northwest of Beijing. The largest habitat of *J. sabina* is found in the Altai Mountains, but the largest habitat in the sandy land is found in the Mu Us Sandy Land (Wang et al. 2011a).

The current area of the *J. sabina* community is less than 1% of the Mu Us Sandy Land, but it is believed that it was distributed over a wider range before anthropogenic desertification occurred (Yoshikawa et al. 1988). In this region, *J. sabina* spreads its stems to crawl on the ground surface, and the community spreads radially. Its canopy grows densely, and mature communities completely cover the ground surface, preventing the movement of sand by wind. Therefore, planting *J. sabina* is one of the important measures in efforts to regenerate original plant communities for desertification prevention and mitigation (Wang et al. 2011a). Ohte et al. (2021) have organized the life history of this species, focusing on natural regeneration by seed reproduction and the growth process of *J. sabina* in relation to the influence of moving sand (see Part IV, Chap. 8).

Besides *J. sabina,* communities of plants of the *Artemisia*, including the semishrub *A. ordosica* of the Asteraceae family, are widespread, and as a shrub of the Fabaceae family, *Caragana korshinskii* forms a sparse community. There are also communities composed of the shrub *Corethrodendron* of the Fabaceae family. These *Artemisia*, *Caragana*, *Corethrodendron*, etc., appear widely in fixed dunes, but also appear well in semi-fixed dunes where a little sand movement occurs. These *Artemisia, Caragana, Corethrodendron*, etc., not only appear widely in fixed dunes but also appear well in semi-fixed dunes where a little sand movement occurs.

**Fig. 5.2** (**a**) Dense stands of *Juniperus sabina* cover the dune surface like carpets. (**b**) Fruits of *J. sabina.* (**c**) Adventitious roots develop from creeping branches

## 5.3 Plants on Semi-fixed Dunes

In semi-fixed dune areas, the coverage of the dune surface by vegetation decreases, and there are many parts where the surface of the sand is exposed. When a strong wind blows, sand is blown away from the bare ground, so the undulations of the ground surface are larger than those of fixed dunes. As of 1993, it accounted for 21.44% of the Mu Us Sandy Land (Wu and Ci 2002).

**Fig. 5.3** (**a**) *Artemisia* colonies grow on semi-fixed dunes (Photo by K. Yoshikawa, Okayama Univ.). (**b**) *Artemisia ordosica* is one of the dominant semi-shrub species found on semi-fixed dunes. (Photo by N. Miki, Okayama Univ.)

In semi-fixed dune areas, like fixed dune areas, the semi-shrub *A. ordosica* and *Artemisia sphaerocephala*, members of the Asteraceae family, form large colonies (Fig. 5.3). *A. ordosica* is widespread in fixed dune areas, while *A. sphaerocephala* is more common in semi-fixed dunes. These species can also colonize bare dunes, making them effective in stabilizing sand dunes. They are also used medicinally.

Also, as a characteristic herbaceous plant in the Mu Us Sandy Land, *Cynanchum mongolicum* sometimes forms colonies in semi-fixed dune areas. *C. mongolicum* is herbaceous plant of the Asclepiadaceae family, which is strong against sand accumulation and can grow above ground even when buried in sand. It is a poisonous plant with alkaloid toxins in its body, and livestock do not eat it. Therefore, in areas with strong grazing pressure, this plant often dominates, making it a good indicator of grazing pressure (Fig. 5.4).

Among woody plants, *Corethrodendron lignosum* var. *laeve* (synonym: *Hedysarum leave*) and *Corethrodendron scoparium* (synonym: *Hedysarum scoparium*) grow mainly in semi-fixed sand dunes. Both are leguminous shrubs. *C. scoparium* is larger than *C. lignosum* var. *laeve* and can reach a height of 2–3 m. The trunk is yellow to brown, and it is characterized by its thin peeling bark. Both species bloom beautiful pink flowers in August, coloring the sand dunes. Their leaves and twigs are important fodder for livestock, and they grow well even in sandy areas. After stopping the movement of sand with grass squares on the sand dunes, *Corethrodendron* seeds of both species are often sown (Fig. 5.5a, b). Also a shrub, *Caragana korshinskii*, member of Fabaceae family, grows mainly in semi-fixed sand dunes and can reach a height of 1–4 m. It shows vigorous growth even when buried in sand, so it is often used as a greening tree species for sand dune fixation. The branches have thorns and are painful to touch, but livestock eat them without concern. In places where grazing is practiced, most of the branch tips are eaten by livestock. It blooms beautiful yellow flowers in spring (Fig. 5.5c, d).

**Fig. 5.4** (**a**) *Cynanchum mongolicum* growing on semi-fixed dunes (Photo by K. Yoshikawa, Okayama Univ.). (**b**) The lower parts of the stems are buried in sand

**Fig. 5.5** *Corethrodendron lignosum* var. *laeve* (**a**) and *Caragana korshinskii* (**c**) growing on semi-fixed dunes. Flowers of *C. lignosum var. laeve* (**b**) and *C. korshinskii* (**d**)

## 5.4 Plants on Shifting Dunes

As of 1993, the area of shifting dunes accounted for 44.53% of the Mu Us Sandy Land (Wu and Ci 2002). In areas of shifting dunes, plants hardly grow, and the entire sand is exposed. Therefore, the movement of sand by wind easily occurs, and the entire dune also moves by wind. Individual shifting dunes are often crescent-shaped dunes (barchan type dunes), but sometimes crescent-shaped sand dunes are connected in a chain. Due to the northwest seasonal winds, shifting dunes often move from northwest to southeast. The movement of sand on the surface of the dune makes it difficult for plants to establish, and only a very small number of psammophytes (plants growing sandy habitat) with strong resistance to sand movement sparsely grow. The characteristic psammophytes in these shifting dunes are the *Agriophyllum squarrosum* (Amaranthaceae) and the *Psammochloa villosa* (Poaceae).

*A. squarrosum* is an annual plant that grows in parts of the shifting dunes where sand movement is relatively small (Fig. 5.6a, b). In years with a lot of rain, many seeds buried in the sand germinate and form large colonies, and if conditions are favorable, individuals can grow over 50 cm tall. As an annual plant, the plant body withers at the end of summer, but the seeds mature and fall on the dunes, becoming buried seeds, and germinate and grow after the rain the following year. Livestock enjoy eating young leaves and stems. The seeds are small, but herders sometimes use them as food.

**Fig. 5.6** Plants growing on shifting sand dunes. *Agriophyllum squarrosum* (**a**, **b**) and *Psammochloa villosa* (**c**, **d**). (Photos by K. Yoshikawa, Okayama Univ.: **a**, **c**)

*P. villosa* is a perennial grass that grows on shifting and semi-fixed dunes. This plant has widely developed rhizomes and forms large colonies through clonal reproduction. It has strong resistance to sand accumulation, and even if it is buried in sand, the stems and leaves come out to the surface. Therefore, it is used as a sand dune fixing plant. The leaves and stems make good fodder for livestock (Fig. 5.6c, d).

## 5.5 Plants of the Interdunal Lowlands

In the Mu Us Sandy Land, the groundwater level is high and the lowland soil is rich in moisture. Therefore, many herbaceous plants grow in the lowlands between the dunes, making it a good grazing area for sheep and goats. A variety of plants grow in response to slight differences in site conditions, and species diversity is high. Also, because salt accumulation is often observed, many of the plants growing in the interdunal lowlands are salt-tolerant.

In the interdunal lowlands, plants grow depending on groundwater, so that high productivity grasslands can be maintained without relying on irregular rainfall. The grasslands that develop in the interdunal lowlands are flat and dominated by perennial plants of the Poaceae and Cyperaceae families. In some places, vegetation composed of broad-leaved perennials can be seen, and they can become beautiful flower fields during the flowering season (Fig. 5.7a, b).

A prominent herbaceous plant in the interdunal lowlands is *Achnatherum splendens*. *A. splendens* is a large perennial grass of the Poaceae family, and the larger ones can reach a height of 1.5 m. It is highly salt tolerant and makes good fodder for livestock.

Also, in the interdunal lowlands, communities of the woody plant, *Salix cheilophila* often develop (Fig. 5.8).

**Fig. 5.7** (**a**) Vegetation dominated by Poaceae and Cyperaceae species in interdunal lowlands. (**b**) Vegetation dominated by broad-leaved perennials. Plants with white flowers are *Leontopodium* sp. (Photos by N. Miki, Okayama Univ.: (**a**), by A. Nakashima, Wakayama Univ.: (**b**))

**Fig. 5.8** (**a**) *Salix cheilophila* growing in interdunal lowland. (**b**): *S. cheilophila* exhibits a tree form consisting of multiple trunks

*S. cheilophila* is a shrubby willow that grows mainly in interdunal lowlands where the groundwater level is high and sufficient soil moisture can be secured. The tree height is usually 1–2 m, and the larger ones can reach 5 m. It has a clump-forming tree shape, and the density of branches and leaves is high, which has the effect of capturing fallen leaves and seeds brought by the wind at the base of the tree. The seeds of *J. sabina* captured at the base of the tree germinate under the canopy, so the harsh environment of high temperature and strong sunlight is mitigated, ensuring a good environment for germination and subsequent growth. Therefore, under the canopy of *S. cheilophila*, there are many seedlings of *J. sabina*, and *S. cheilophila* plays an important role in the regeneration of *J. sabina*.

The state of the groundwater level in the interdunal lowlands varies greatly from year to year, with some lowlands always having water and others being completely dry. In places where water accumulates for a long time, aquatic plants such as members of the genus *Typha* can grow.

Soil salinization and alkalization increase when water mobility is low and stagnant, forming saline soil (with a salt content of 0.3% or more) and alkaline soil (with a pH of 8–10). In particular, in the vast grassland areas that develop in lowlands, small-grained clayey soils are distributed instead of sand and, therefore, there is a tendency for salinization and alkalization to progress strongly. (Chen 1983).

In the southern edge of the Mu Us Sandy Land where salt accumulation is notably observed, shrub communities dominated by halophytic shrubs such as *Tamarix austromongolica* and *Tamarix chinensis* develop (Fig. 5.9). In such halophytic plant communities, in addition to *Tamarix* spp., many halophytic plants including shrubs such as *Suaeda, Kalidium,* and *Nitraria* grow.

**Fig. 5.9** *Tamarix austromongolica* growing in salt-accumulated areas

## References

Chen C (1983) Vegetation and plant resources. In: the Department of Geography of Beijing University et al (eds) Natural conditions and their improvement and utilization in the Mu Us Sandy land. Science Press, Beijing, pp 72–113 (in Chinese)

Kobayashi T (1994) Studies on ecology and water relations of revegetation plants in the Mu-Us sandy district, China. Tech Bull Fac Hort Chiba Univ 48:329–381 (in Japanese)

Ohte N, Miki NH, Matsuo N, Yang L, Hirobe M, Yamanaka N, Ishii Y, Tanaka-Oda A, Shimizu M, Zhang G, Sakamoto K, Wang L, Yoshikawa K (2021) Life history of *Juniperus sabina* L. adapted to the sand shifting environment in the Mu Us Sandy Land, China: a review. Landsc Ecol Eng 17:281–294

Wang L, Zhang G, Wen G, Li Y, Yoshikawa K (2011a) Physiological and ecological characteristics of *Juniperus sabina,* and population recovery and reconstruction. Science Press, Beijing (in Chinese)

Wang T et al (2011b) Deserts and aeolian desertification in China. Science Press, Beijing

Wu B, Ci L (2002) Landscape change and desertification development in the Mu Us Sandland, Northern China. J Arid Environ 50:429–444

Yoshikawa K, Li S, Kobashi S, Masuda T (1988) A preliminary survey of plant biomass at the Mu Us desert in Inner Mongolia. Bull Reveget Res 10:66–102 (in Japanese)

# Part III
# Physiological and Ecological Characteristics of Plants Under Drought and Strong Light Stress

## Introduction

In addition to being an environment with limited water resources, drylands are under harsh environments in various conditions such as light and temperature. How do plants in drylands survive such harsh environments and with what mechanisms, characteristics, and strategies?

In this part, we will introduce the mechanisms and characteristics of plants in the Mu Us Sandy Land against drought stress and strong light stress. The mechanisms to drought stress and strong light stress have many related, and it can be the core mechanisms for surviving in drylands as there are also parts related to other stresses such as temperature.

# Chapter 6
# Mechanisms Against Drought Stress

**Naoko H. Miki and Guosheng Zhang**

## 6.1 Effect of Drought on Plants

Drylands are environments where the availability of water for plants is extremely low because the amount of precipitation is small and the potential evapotranspiration exceeds the amount of precipitation. Plants acquire carbon through the stomata of leaves to perform photosynthesis, but at the same time, they lose water through transpiration through the stomata. When water is lost from the leaves due to transpiration, the water absorbed by the roots is transported to the leaves through the stem. However, under drought, the water balance in the plant is disrupted because the absorption and transport of water cannot keep up with the loss of water, resulting in drought stress. When plants are subjected to drought stress, a decrease in turgor pressure occurs, leading to wilting and death. In addition, metabolic activity in the protoplasm deteriorates and stomata close, which suppresses photosynthesis and inhibits growth. Therefore, mechanisms to drought stress are important for growth and survival in drylands.

## 6.2 Mechanisms Against Drought Stress

Plants that grow in dry conditions and have drought resistance, the capacity of plants to withstand periods of dryness, are called xerophytes. Xerophytes have drought resistance mechanisms such as "desiccation postponement," "desiccation tolerance," and "drought escape" to survive in dry environments (Larcher 2003; Taiz and Zeiger 2006).



K. Yoshikawa et al., *Dryland Vegetation in the Sand-Shifting Environment*,
https://doi.org/10.1007/978-981-95-4735-7_6

### 6.2.1 Desiccation Postponement

Many plant species in drylands have various mechanisms to suppress or delay the reduction of the plant-water status due to transpiration (water deficit), such as suppressing water loss from leaves, immediately absorbing water at the roots in response to water loss, and maintaining water transport function under various water conditions in the stem. These mechanisms are interrelated within the whole-plant.

Water flows through the stem due to the tension (negative pressure) caused by transpiration. However, if transpiration continues while water absorption does not proceed well due to drought, and the tension in the water transport tissue of xylem increases and exceeds a certain threshold, an air bubble is aspirated from air-filled conduit to water-filled conduit via pores in the inter-conduit pit membranes and expands (cavitation), blocking water flow in the conduit as embolism (Fig. 6.1) (Sperry and Tyree 1988; Jansen et al. 2009). When cavitation occurs, the water transport function decrease. So for plants growing in drylands, it is important to avoid the occurrence of cavitation by increasing resistance to cavitation and by alleviating the tension in the xylem due to suppress of transpiration and efficient water absorption so as not to lose the water transport function as much as possible under dry conditions.

Regarding the plants of the Mu Us Sandy Land, the prostrate evergreen conifer *Juniperus sabina* (Fig. 6.2) has remarkably higher xylem cavitation resistance than other broad-leaved woody species such as *Artemisia ordosica*, *Artemisia sphaerocephala*, *Caragana microphylla*, *Salix psammophila*, *Salix cheilophila* (Fig. 6.3, Murakami et al. 2016). Also, *J. sabina* has a low stomatal density (Dong and Zhang

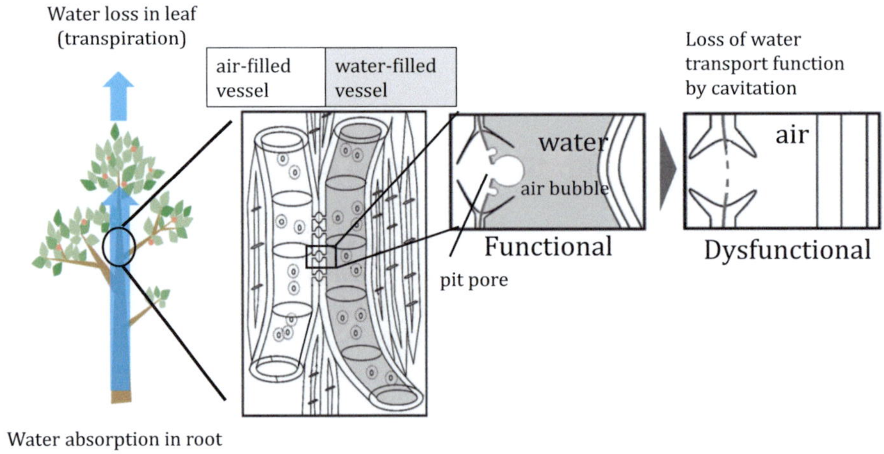


**Fig. 6.1** Excessive tension caused by transpiration lead to cavitation of conduit and loss of water transport function. (Modified from Jacobsen et al. 2005)

**Fig. 6.2** *J. sabina* has prostrate stems and forms a dense patch

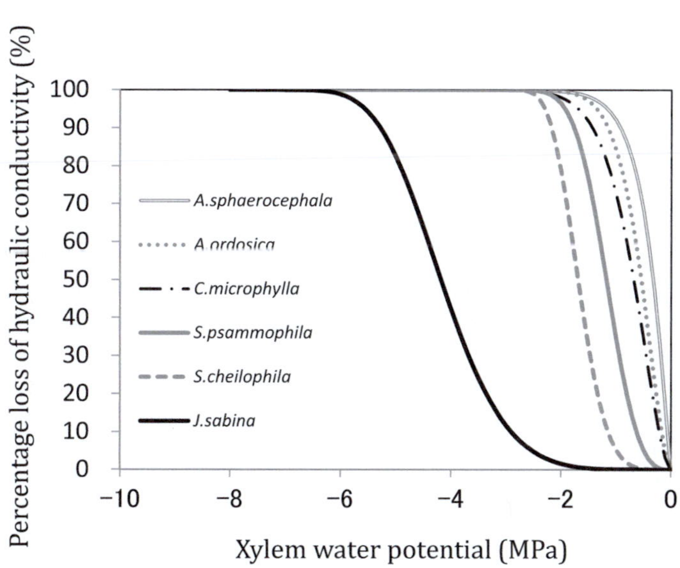


**Fig. 6.3** Curves for loss of xylem water transport function due to dehydration. (Modified from Murakami et al. 2016)

**Table 6.1** Mean (±SE) stomatal conductance during midday (9:00,10:00,13:00) *for J. sabina, A. ordosica,* and *C. korshinskii*

| | Stomatal conductance (mol $m^{-2}$ $s^{-1}$) |
|---|---|
| *J. sabina* | 0.048 ± 0.018 |
| *A. ordosica* | 0.188 ± 0.072 |
| *C. korshinskii* | 0.449 ± 0.132 |

Modified from Matsuo et al. (2001)

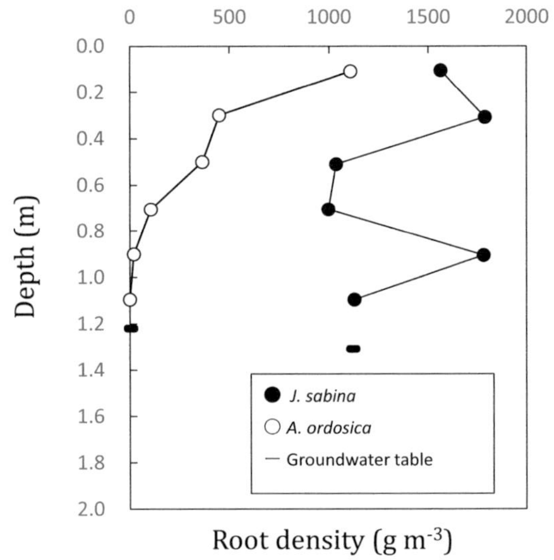


**Fig. 6.4** Profiles of root density for *J. sabina* and *A. ordosica*. (Modified from Ohte et al. 2003)

2000), and compared to other broad-leaved woody species, such as *A. ordosica* and *Caragana korshinskii*, its transpiration rate is markedly lower and water loss is suppressed (Table 6.1, Matsuo et al. 2001). Regarding root water absorption, while *A. ordosica* is a plant that uses rainfall by spreading its roots in the shallow soil layer near the surface, *J. sabina* has adventitious roots in addition to the main root and spreads its roots at various depths from the ground surface to near the groundwater level (Fig. 6.4, Zhang et al. 1999; Ohte et al. 2003). Some plants in drylands that have root systems in different soil layers redistribute water through their roots between different soil layers (hydraulic redistribution), achieving efficient water use (Caldwell et al. 1998; Loik 2007). Hydraulic redistribution is the passive movement of water from moist to dry soil through plant roots (Fig. 6.5) (Neumann and Cardon 2012). *J. sabina* also redistributes water through its roots from the relatively moist deep soil layers affected by groundwater to the dry soil surface layers at nighttime during periods without rain (Fig. 6.6, Yang et al. 2014; Miki et al. 2016, 2018), ensuring growth by effectively absorbing water in response to occasional slight rain

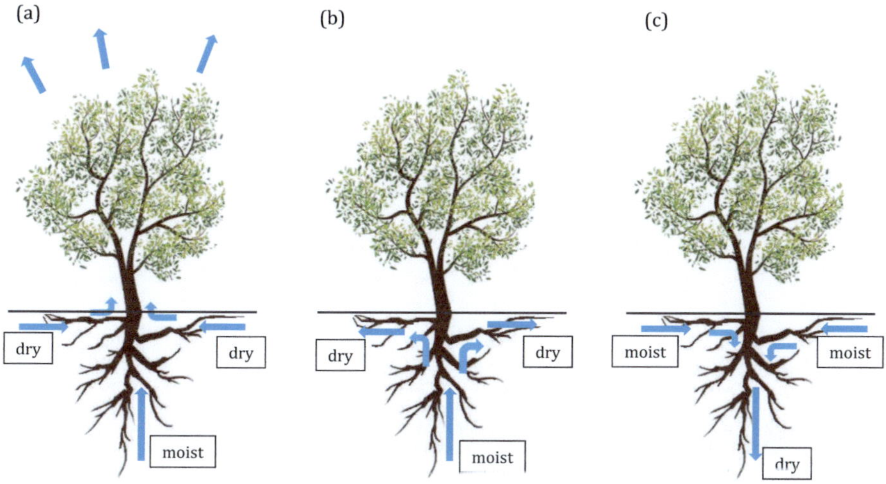


**Fig. 6.5** Passive movement of water via roots. The arrows show the direction of water flow. (**a**) Water flow from root to leaves at daytime, (**b**) water flow from deeper soil layer to surface layer at nighttime during periods without rain (HL hydraulic lift), (**c**) water flow from surface layer to deeper layer at nighttime after rain (HD hydraulic descent). (**b**) and (**c**) show the hydraulic redistribution

in addition to groundwater in environments where water availability is limited (Fig. 6.7, Miki et al. 2016). Therefore, it is thought that *J. sabina* maintains its water transport function by high resistance to xylem cavitation, suppression of leaf water loss, and alleviation of xylem tension through effective root water absorption due to hydraulic redistribution.

On the other hand, other broad-leaved woody species have lower xylem cavitation resistance than *J. sabina* and are more likely to lose their water transport function (Fig. 6.3, Murakami et al. 2016). When subjected to drought stress, cavitation occurs and water transport function decreases, but when drought stress is relieved by rainfall, some species of broad-leaved woody species, which mainly have vessels as water transport tissues, can recover their water transport function in the short term (Salleo et al. 1996; Ogasa et al. 2010, 2013). This short-term recovery is thought to involve the refilling of cavitated vessels with water (Jacobsen et al. 2015). An evaluation of the recovery of water transport function in species of the genus *Salix* showed a tendency to recover (Fig. 6.8), so species of the genus *Salix*, which have particularly low cavitation resistance among broad-leaved woody species, are thought to maintain their water transport function under fluctuating water conditions by compensating for the risk of loss of water transport function due to low cavitation resistance with high recovery (Murakami et al. 2016, Ogasa et al. 2013).

Also, when these species of the Mu Us Sandy Land were compared with other species in semi-arid area and species in humid area, the species of the Mu Us Sandy Land, both the conifer *J. sabina* and the five broad-leaved woody species, have

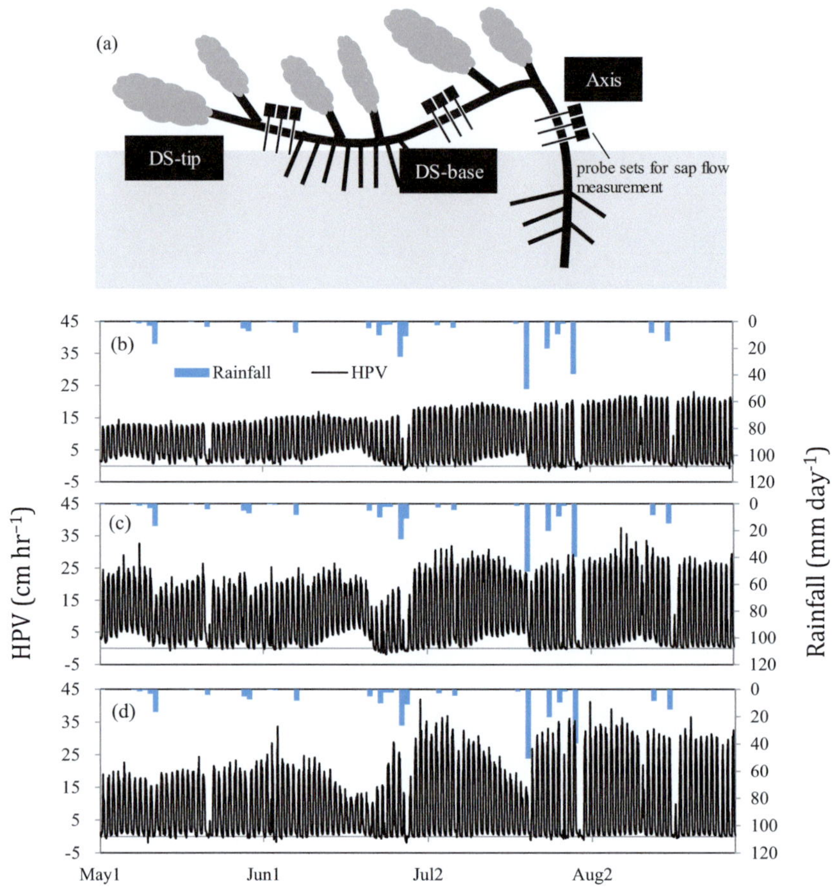


**Fig. 6.6** (**a**) Installation position of probe sets for sap flow measurement in studied individual and changes in rainfall and heat pulse velocity (HPV) in (**b**) axis, (**c**) DS-base, and (**d**) DS-tip (Modified from Miki et al. 2016). Nighttime sap flow in the axis and DS-base appeared during periods without rainfall and it decreased or disappeared after a rainfall event (**b**, **c**), while nighttime sap flow was not observed in the DS-tip even during dry soil conditions (**d**). These nighttime sap flows have verified as hydraulic redistribution by tracer experiments with $^{18}$O-labeled water. (See Miki et al. 2018)

lower cavitation resistance and smaller safety against xylem cavitation, represented by the difference between the minimum xylem water potential during summer midday and cavitation resistance (Ψ50; xylem water potential causing a 50% loss in xylem transport function), than other species in semi-arid environments, and tend to be as prone to cavitation as species in humid environments (Fig. 6.9, Murakami et al. 2016). The Mu Us Sandy Land is located in a shifting sand environment, and many of the species growing in the Mu Us Sandy Land produce adventitious roots (Fig. 6.10) when their branches and/or stems are buried in sand due to wind-blown sand. Therefore, in addition to being located under conditions with relatively high groundwater level in semi-arid area (see Sect. 4.3.1), species in the Mu Us Sandy

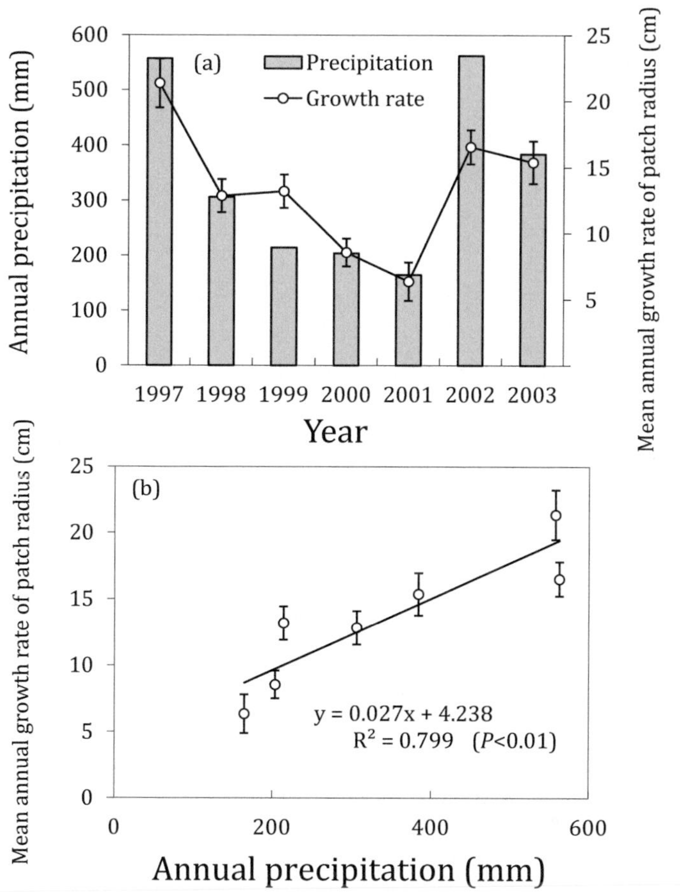


**Fig. 6.7** (**a**) Annual precipitation and mean annual growth rate of *J. sabina* patch radius (mean ± SE) and (**b**) the correlation between annual precipitation and mean annual growth rate of patch radius. (Reproduced from Miki et al. 2016)

Land distribute their roots widely and ensure water absorption, enabling the avoidance or recovery of water transport function loss and compensating for their low cavitation resistance.

### 6.2.2 *Desiccation Tolerance*

Many vascular plants have a mechanism to maintain intracellular pressure (turgor pressure) and survive even under dehydration conditions. By accumulating solutes such as sugars inside cells, they prevent plasmolysis and loss of turgor pressure due

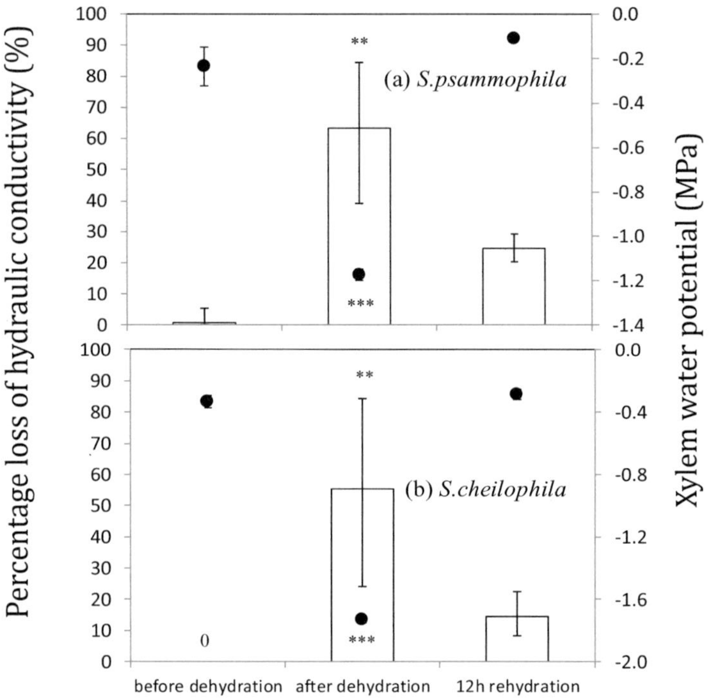


**Fig. 6.8** Percentage loss of xylem water transport function before dehydration, after a 12-h dehydration and after rehydration period for *Salix* species. (Reproduced from Murakami et al. 2016)

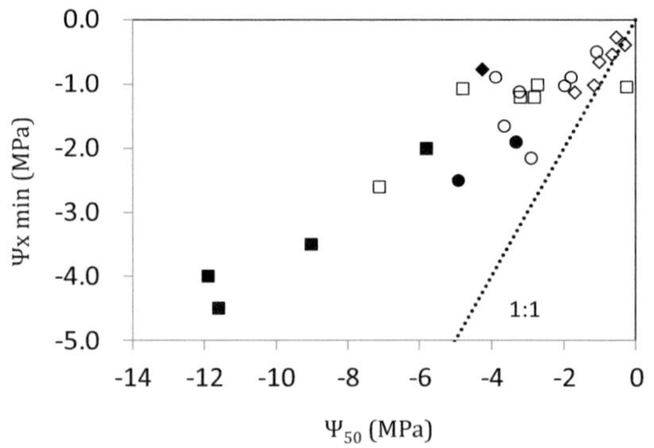


**Fig. 6.9** Relationship between midday xylem water potential (Ψxmin) and xylem water potential causing a 50% loss in xylem transport function (Ψ50; cavitation resistance) (Reproduced from Murakami et al. 2016). Filled squares, conifers in a semi-arid environment; open squares, broadleaf trees in a semi-arid environment; filled circles, conifers in a humid environment; open circles, broadleaf trees in a humid environment; filled diamonds, conifers in the Mu Us Sandy Land; open diamonds, broadleaf trees in the Mu Us Sandy Land

**Fig. 6.10** Adventitious roots of the species growing in the Mu Us Sandy Land. (**a**) *J. sabina*, (**b**) *A. ordosica* (Photo by R. Sasaki, Okayama Univ.), (**c**) *C. mongolicum*, (**d**) *S. psammophila*. (Photo by A. Nakashima, Wakayama Univ.)

to dehydration (osmotic adjustment), maintain metabolic activity against drying, and can prevent wilting and death.

The main physiological traits of leaves, estimated from the relationship between leaf water potential and relative leaf water content known as the pressure-volume curve, include leaf water potential at turgor loss point (ψwtlp), relative water content at turgor loss point (RWCtlp), and leaf osmotic potential at full turgor (ψo). Among these, ψwtlp is the trait that most directly quantifies a plant drought tolerance (Bartlett et al. 2014) and is the negative water potential when leaf cells lose turgor and leaves wilt, closing stomata and ceasing gas exchange and growth (Cheung et al. 1975; Brodribb et al. 2003; Blackman et al. 2010). Therefore, the more negative the value of ψwtlp, the more gas exchange such as photosynthesis can be maintained under dry soil conditions (Brodribb et al. 2003; Bartlett et al. 2012). The value of ψo depends on the changes in the solute concentration of the symplast due to osmotic adjustment, and a lower ψo tends to result in a lower ψwtlp (Cheung et al. 1975; Bartlett et al. 2012).

For plants in the Mu Us Sandy Land, when comparing the evergreen conifer *J. sabina* (Cupressaceae), the deciduous semi-shrub *A. ordosica* (Asteraceae), the

**Table 6.2** Leaf water relations characteristics derived from pressure–volume curve analysis

| | Ψwtlp (MPa) | Ψo (MPa) | RWCtlp (%) |
|---|---|---|---|
| *J. sabina* | −2.38 | −1.29 | 68.5 |
| *A. ordosica* | −1.36 | −1.19 | 78.8 |
| *C. korshinskii* | −1.02 | −0.83 | 84.5 |
| *S. psammophila* | −1.22 | −1.08 | 92.7 |
| *S. cheilophila* | −1.25 | −1.09 | 92.8 |

Modified from Yoshikawa et al. (1992), Yan et al. (2017)

deciduous shrub *C. korshinskii* (Fabaceae), and the deciduous shrub or small tree *S. psammophila* and *S. cheilophila* (Salicaceae), the ψo representing osmotic adjustment ability is in the order of *J. sabina*, *A. ordosia*, *S. cheilophila*, *S. psammophila*, *C. korshinskii* from low to high. And, ψwtlp is also in the order of *J. sabina* at −2.38 MPa, followed by *A. ordosia* at −1.36 MPa, *S. cheilophila* at −1.25 MPa, *S. psammophila* at −1.22 MPa, and *C. korshinskii* at −1.02 MPa (Table 6.2, Yoshikawa et al. 1992; Yan et al. 2017). Among the species growing in the Mu Us Sandy Land, *J. sabina* has an excellent mechanism for desiccation postponement (see Sect. 6.2.1), such as suppressing water loss in leaves, effective water absorption in roots, and maintaining water transport function under various water conditions in the stem, and, in addition, desiccation tolerance due to osmotic adjustment is higher than other species.

On the other hand, when looking at the ψwtlp for each aridity index (AI; the ratio of mean annual precipitation to mean annual potential evapotranspiration) (Zhu et al. 2018), species in the Mu Us Sandy Land generally tend to have relatively high ψwtlp even among species in semi-arid area (AI from 0.2 to 0.5; UNEP 1997; Millennium Ecosystem Assessment 2005), and some species show values close to those of species in humid environments. This may be influenced by the fact that it is located under conditions with relatively high groundwater level even in semi-arid area (see Sect. 4.3.1), and that species growing in the Mu Us Sandy Land can ensure water absorption even under fluctuating soil moisture conditions due to effective water absorption by adventitious roots and main roots.

### *6.2.3 Drought Escape*

To prevent damage to their leaves and flowers from drought, many short-lived annual herbs overcome the dry period mainly through their seeds, germinating, growing, flowering, and fruiting during the brief moist period after rainfall, thus completing their life cycle. On the other hand, perennial herbs spend the dry period with drought-resistant dormant buds, forming leaves and flowers all at once using the subsequent rainfall and bearing fruit. They have developed phenology that can avoid desiccation, or have developed drought-resistant seeds or dormant buds to avoid the dry season itself.

In the Mu Us Sandy Land, there are few species that create a field of flowers like the so-called desert ephemerals, but many annual and perennial herbs are distributed. For example, among the annual herbs, there are *Cynanchum mongolicum* from the Asclepiadaceae family, *Agriophyllum squarrosum* from the Chenopodiaceae family, and among the perennial herbs, there are *Stipa breviflora*, *Stipa bungeana*, *Psammochloa villosa*, and *Achnatherum splendens* from the Poaceae family (Liu and Wang 2006; Zhao et al. 2019).

## References

Bartlett MK, Scoffoni C, Sack L (2012) The determinants of leaf turgor loss point and prediction of drought tolerance of species and biomes: a global meta-analysis. Ecol Lett 15:393–405

Bartlett MK, Zhang Y, Kreidler N, Sun S, Ardy R, Cao K, Sack L (2014) Global analysis of plasticity in turgor loss point, a key drought tolerance trait. Ecol Lett 17:1580–1590

Blackman CJ, Brodribb TJ, Jordan GJ (2010) Leaf hydraulic vulnerability is related to conduit dimensions and drought resistance across a diverse range of woody angiosperms. New Phytol 188:1113–1123

Brodribb T, Holbrook NM, Edwards EJ, Gutierrez MV (2003) Relations between stomatal closure, leaf turgor and xylem vulnerability in eight tropical dry forest trees. Plant Cell Environ 26:443–450

Caldwell MM, Dawson TE, Richards JH (1998) Hydraulic lift: consequences of water efflux from the roots of plants. Oecologia 113:151–161

Cheung YNS, Tyree MT, Dainty J (1975) Water relations parameters on single leaves obtained in a pressure bomb and some ecological interpretations. Can J Bot 53:1342–1346

Dong X, Zhang X (2000) Special stomatal distribution in Sabina vulgaris in relation to its survival in a desert environment. Trees 14:369–375

Jacobsen AL, Ewers FW, Pratt RB, Paddock WA, Davis SD (2005) Do xylem fibers affect vessel cavitation resistance? Plant Physiol 139(1):546–556

Jacobsen AL, Rodriguez-Zaccaro FD, Lee TF, Valdovinos J, Toschi HS, Martinez JA, Pratt RB (2015) Grapevine xylem development, architecture, and function. In: Hacke UG (ed) Functional and ecological xylem anatomy. Springer, New York, pp 133–162

Jansen S, Choat B, Pletsers A (2009) Morphological variation of intervessel pit membranes and implications to xylem function in angiosperms. Am J Bot 96:409–419

Larcher W (2003) Physiological plant ecology: ecophysiology and stress physiology of functional groups, 4th edn. Springer, p 514

Liu XQ, Wang RZ (2006) Photosynthetic pathway and morphological functional types in the vegetation from North-Beijing agro-pastoral ecotone, China. Photosynthetica 44(3):365–386

Loik ME (2007) Sensitivity of water relations and photosynthesis to summer precipitation pulses for *Artemisia tridentata* and *Purshia tridentate*. Plant Ecol 191:95–108

Matsuo N, Ohte N, Koba K, Kosugi Y, Kabeya N, Zhang GS, Wang LH, Yoshikawa K (2001) Analysis of water-use efficiency of plants in the Mu-Us desert, Inner Mongolia, China using carbon isotope discrimination. J Jpn Soc Reveget Tech 27(1):68–73 (in Japanese with English summary)

Miki NH, Kubori F, Yang L, Matsuo N, Zhang G, Wang L, Yoshikawa K (2016) Effective water use and growth of a prostrate lifeform shrub, Juniperus sabina, in semiarid areas of China. J Arid Land Stud 26:95–99

Miki NH, Sato K, Aoki M, Yang L, Matsuo N, Zhang G, Wang L, Yoshikawa K (2018) Water movement via adventitious roots of the prostrate shrub Juniperus sabina in semiarid areas of China. Acta Hortic 1222:137–146

Millennium Ecosystem Assessment (2005) Ecosystems and human well-being: current state and trends: chapter 22 dryland systems. Island Press, Washington, DC, pp 623–662
Murakami Y, Miki NH, Yang L, Zhang G, Wang LH, Yoshikawa K (2016) Water transport properties of seven woody species from the semi-arid Mu Us Sandy Land, China. Landsc Ecol Eng 12:209–220
Neumann RB, Cardon ZG (2012) The magnitude of hydraulic redistribution by plant roots: a review and synthesis of empirical and modeling studies. New Phytol 194(2):337–352
Ogasa M, Miki N, Yoshikawa K (2010) Changes of hydraulic conductivity during dehydration and rehydration in *Quercus serrata* Thunb. and *Betula platyphylla* var. *japonica* Hara: the effect of xylem structures. Tree Physiol 30:608–617
Ogasa M, Miki NH, Murakami Y, Yoshikawa K (2013) Recovery performance in xylem hydraulic conductivity is correlated with cavitation resistance for temperate deciduous tree species. Tree Physiol 33:335–344
Ohte N, Koba K, Yoshikawa K, Sugimoto A, Matsuo N, Kabeya N, Wang L (2003) Water utilization of natural and planted trees in the semiarid desert of Inner Mongolia. China Ecol Appl 13(2):337–351
Salleo S, Lo Gullo MA, De Paoli D, Zippo M (1996) Xylem recovery from cavitation-induced embolism in young plants of Laurus nobilis: a possible mechanism. New Phytol 132:47–56
Sperry JS, Tyree MT (1988) Mechanism of water stress-induced xylem embolism. Plant Physiol 88:581–587
Taiz L, Zeiger E (2006) Plant physiology, 4th edn. Sinauer Associate Inc, p 764
UNEP (1997) World atlas of desertification, 2nd edn. Arnold, p 192
Yan Z, Zhao X, Hao L, W Y, Bai Y, Zhang G, Li X (2017) Drought resistance evaluation of 16 plants based on pressure-volume curve water parameters. J West China For Sci 46(1):112–116 (in Chinese with English summary)
Yang L, Miki NH, Matsuo N, Zhang G, Wang L, Yoshikawa K (2014) Contribution of adventitious roots to water use strategy of Juniperus sabina in a semiarid area of China. J Agric Sci Technol A 4(3):251–259
Yoshikawa K, Kobayashi T, Kobashi S (1992) Tissue water relations characteristics of tree species in semi arid region in China. J Jpn Soc Reveget Tech 17(2):85–93 (in Japanese with English summary)
Zhang G, Wang L, Li Y, Dong Z, Qi J (1999) Root distribution and root weight of Sabina vulgaris in Mu Us Sandy Land. J Desert Res 19(4):378–383 (in Chinese with English summary)
Zhao YD, He XD, Chen L, Ding XF, Li MQ, Xu PY, Gao YB (2019) Features on N/P ratio of plants with different functional groups between two types of steppe in semi-arid area. Sci Cold Arid Reg 11(5):371–381
Zhu SD, Chen YJ, Ye Q, He PC, Liu H, Li R, Fu PL, Jiang GF, Cao KF (2018) Leaf turgor loss point is correlated with drought tolerance and leaf carbon economics traits. Tree Physiol 38:658–663

# Chapter 7
# Mechanisms Against Strong Light Stress

**Naoko H. Miki and Guosheng Zhang**

## 7.1 Effect of Strong Light on Plants

The light energy absorbed by the leaves is used for photosynthesis. However, under strong light above the light saturation point, there is a shortage of carbon dioxide relative to the amount of light energy, so not all the absorbed light energy can be used for photosynthesis, resulting in excess light energy (strong light stress). As a result, harmful reactive oxygen species (ROS) are generated, the photosynthetic system is damaged, and photosynthesis is inhibited (photoinhibition) (Fig. 7.1, Taiz and Zeiger 2006).

When subjected to drought stress, the stomata close to suppress transpiration, so carbon dioxide cannot be taken in, the concentration of carbon dioxide in the leaves decreases, and the light energy that cannot be used in photosynthesis increases even more. Also, the biosynthetic reactions that synthesize organic compounds in photosynthesis are affected by temperature because they involve enzyme-catalyzed reactions. Generally, the optimal temperature for photosynthesis is between 20 and 35 °C, so when the temperature is low extremely, especially in winter, the biosynthetic reactions are suppressed and the light energy that cannot be used in photosynthesis increases.

Drylands are not only under conditions where soil moisture availability is low and drought stress is likely to occur, but also have very high solar radiation, so the dryness of the atmosphere and light intensity are also remarkably high. Furthermore, in high latitude or high altitude regions, the temperature is extremely low in winter, making it an environment where strong light stress is likely to be amplified. Therefore, it is important for plants growing in drylands to have some photoprotective mechanism against strong light stress.



K. Yoshikawa et al., *Dryland Vegetation in the Sand-Shifting Environment*,
https://doi.org/10.1007/978-981-95-4735-7_7

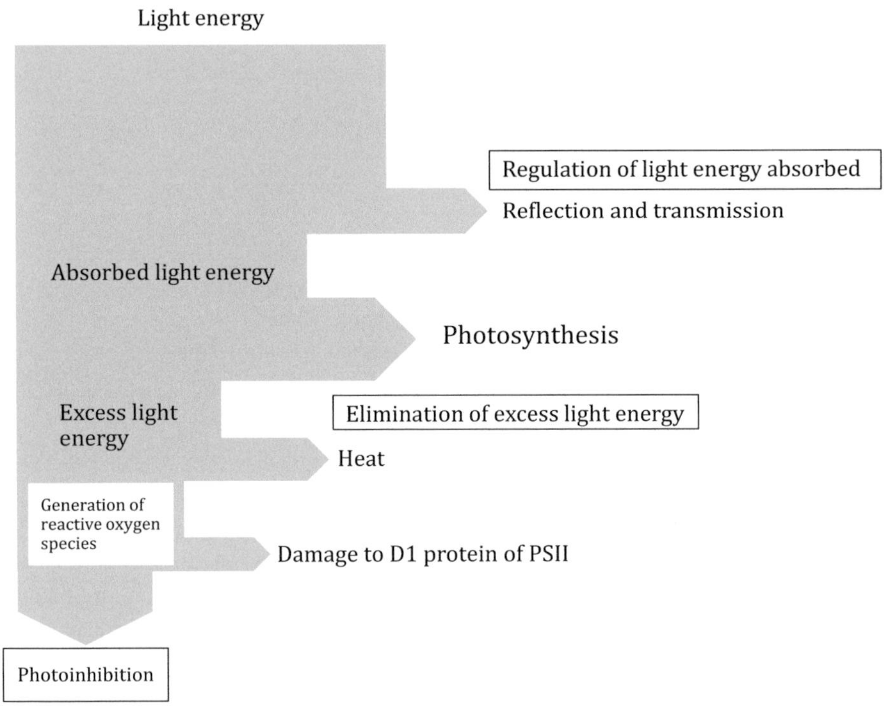


**Fig. 7.1** Flow of light energy. (Modified from Taiz and Zeiger 2006)

## 7.2 Photoprotection Mechanisms Against Strong Light Stress

The light energy absorbed by the leaves is used for photosynthesis, but under strong light, not all the absorbed light energy can be used for photosynthesis, resulting in excess light energy (Fig. 7.1). Therefore, by changing the composition ratio of pigments such as chlorophyll involved in light absorption, the amount of light energy absorbed is suppressed, thereby suppressing the generation of excess light energy (regulation of the amount of light energy absorbed). However, if the absorbed light energy is still too much, the pool size and composition of pigments such as carotenoids are changed to dissipate the excess light energy as heat (heat dissipation) and eliminate it (elimination of absorbed excess light energy). Furthermore, if the excess energy cannot be completely eliminated, specific proteins eliminate the energy in the form of damage. The damaged proteins are then synthesized anew and repaired.

On the other hand, in dry environments where water availability is scarce, it is important to maintain photosynthetic activity and to use as much of the absorbed light energy as possible for photosynthesis in order to suppress the generation of excess light energy (suppression of excess light energy generation). The more the light used for photosynthesis, the more the generation of excess light energy can be suppressed.

One mechanism to suppress the amount of excess light energy in dry environments by utilizing as much of the absorbed light energy as possible for photosynthesis is heterophylly. Heterophylly refers to variation in leaf phenotypes such as size and/or shape and is observed in various plants (Winn 1999). Heterophylly is recognized in many plant species such as the cypress family based on growth stages, growing seasons, and environments, but the degree of heterophylly varies greatly by species, ranging from subtle differences in leaf size to dramatic differences in both leaf size and shape (Mommer and Visser 2005). Leaf morphology greatly affects leaf functions such as photosynthesis and transpiration (Larcher 2003), so leaf heterophylly is thought to play an important role in combating various environmental conditions (Winn 1999).

## 7.3 Photoprotection Mechanism Against Strong Light Stress in *J. sabina*

*Juniperus sabina*, an evergreen shrub of the cypress family, which is one of the dominant species in the Mu Us Sandy Land, is known to exhibit heterophylly (Tanaka-Oda et al. 2010). Seedlings and saplings growing under the coverage of shrub species have needle leaves, and when they become mature individual growing in sunny conditions outside the shrub coverage, they have both needle and scale leaves, with a remarkably higher proportion of scale leaves (Fig. 7.2, Tanaka-Oda et al. 2010).

The needle leaves of *J. sabina* have a low light compensation point and water use efficiency and have the characteristic of efficiently performing photosynthesis while suppressing water loss under weak light conditions (Tanaka-Oda et al. 2010). Therefore, they suppress the generation of excess light energy that cannot be fully used for photosynthesis by having leaves under weak light conditions such as inside the shrub coverage. Even when excess light energy is generated, they change the composition ratio of chlorophyll pigments related to light absorption according to the difference in light intensity where they grow, both in summer and winter, and have a mechanism to suppress the amount of light energy absorbed against strong light (Fig. 7.3, Ishii et al. 2006b). In addition, regulation is observed in the pool size (V + A + Z) and composition ratio ((A + Z)/(V + A + Z)) of xanthophyll cycle pigments (violaxanthin (V), antheraxanthin (A), zeaxanthin (Z)), which are one of carotenoid. As the light intensity increases, the pool size becomes larger, and the composition ratio of pigments involved in heat dissipation (A + Z) is high, indicating that they are eliminating excess light energy by heat dissipation (Fig. 7.4, Ishii et al. 2006b). In winter, photosynthetic activity is remarkably reduced due to low temperatures, and the pool size of xanthophyll cycle pigments and the composition ratio of pigments involved in heat dissipation are remarkably higher than in summer, indicating that they are actively eliminating excess light energy by heat dissipation (Fig. 7.5, Ishii et al. 2006a, b).

**Fig. 7.2** Photos of scale and needle leaves from mature individuals and needle leaves from seedlings for *J. sabina*. (Reproduced from Tanaka-Oda et al. 2010)

**Fig. 7.3** Relationship between growth-relative light intensity and Chl-a/Chl-b ratio in summer (filled circles) and winter (open circles). (Reproduced from Ishii et al. 2006b)

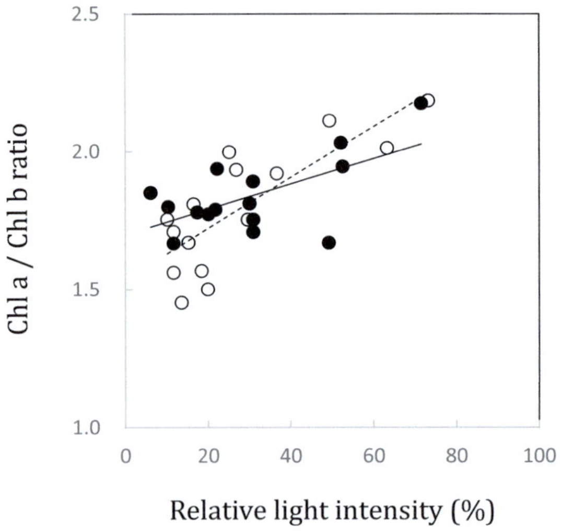

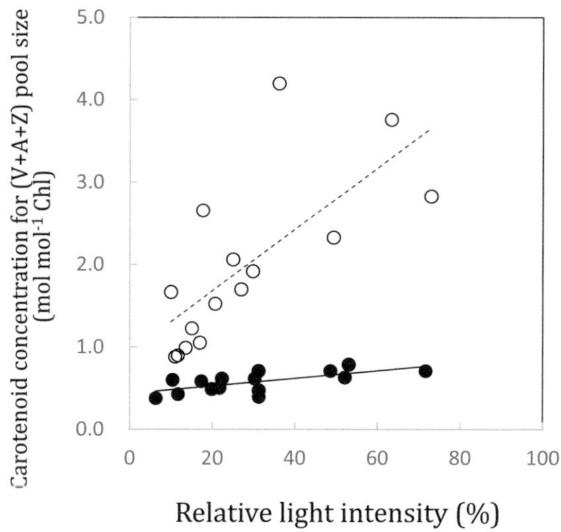


**Fig. 7.4** Relationship between growth-relative light intensity and carotenoid concentrations on a chlorophyll basis for (V + A + Z) pool size in summer (filled circles) and winter (open circles). (Reproduced from Ishii et al. 2006b)

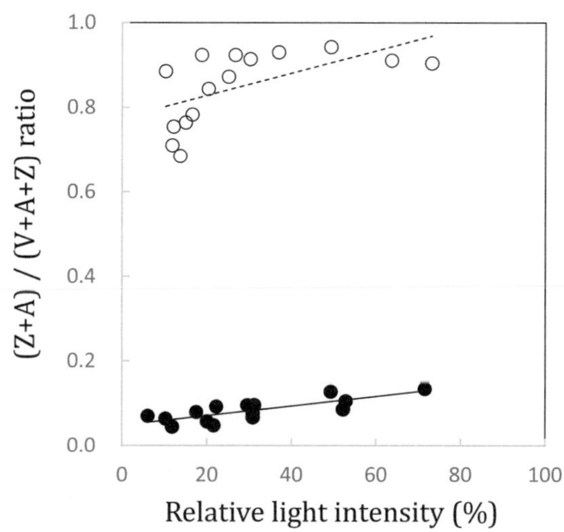


**Fig. 7.5** Relationship between growth-relative light intensity and (Z + A)/(V + A + Z) ratio in summer (filled circles) and winter (open circles). (Reproduced from Ishii et al. 2006b)

On the other hand, scale leaves, compared to needle leaves, have a thicker cuticle layer, lower stomatal density (Dong and Zhang 2000; He and Zhang 2003; He et al. 2003), and a larger LMA, which represents the dry weight per leaf area (Tanaka-Oda et al. 2010). They also have a high photosynthetic rate and water use efficiency under strong light, which effectively suppresses the generation of excess light energy, and they have a high ability to recover the photosystem II damaged by strong light (Fig. 7.6, Tanaka-Oda et al. 2010).

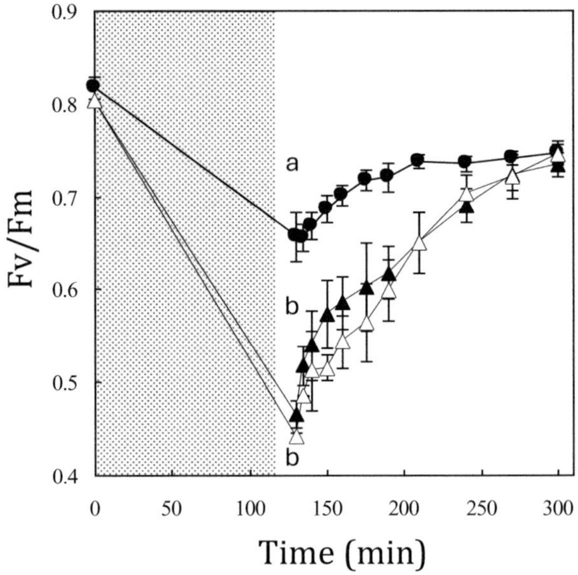


**Fig. 7.6** Susceptibility to photoinhibition of scale leaf (filled circle) and needle leaf (filled triangle) of mature individuals and needle leaf of seedlings (open triangle). Exposure to photoinhibitory light is indicated by dotted area. Different letters indicate significant differences across the leaf types according to Tukey's HSD test after ANOVA. (Reproduced from Tanaka-Oda et al. 2010)

## References


Dong X, Zhang X (2000) Special stomatal distribution in Sabina vulgaris in relation to its survival in a desert environment. Trees 14:369–375

He WM, Zhang XS (2003) Responses of an evergreen shrub Sabina vulgaris to soil water and nutrient shortages in the semiarid Mu Us Sandland in China. J Arid Environ 53:307–316

He WM, Zhang XS, Dong M (2003) Gas exchange, leaf structure, and hydraulic features in relation to sex, shoot form, and leaf form in an evergreen shrub Sabina vulgaris in the semi-arid Mu Us Sandland in China. Photosynthetica 41:105–109

Ishii Y, Sakamoto K, Wang LH, Yoshikawa K (2006a) Seasonal changes in pigment composition of Sabina vulgaris needles in the Mu Us Sandy Land of Northern China. J Arid Land Stud 16(2):85–93

Ishii Y, Sakamoto K, Yamanaka N, Wang L, Yoshikawa K (2006b) Light acclimation of needle pigment composition in Sabina vulgaris seedlings under nurse plant canopy. J Arid Environ 67:403–415

Larcher W (2003) Physiological plant ecology: ecophysiology and stress physiology of functional groups, 4th edn. Springer, p 514

Mommer L, Visser EJW (2005) Underwater photosynthesis in hooded terrestrial plants: a matter of leaf plasticity. Ann Bot 96:581–589

Taiz L, Zeiger E (2006) Plant physiology, 4th edn. Sinauer Associate Inc, p 764

Tanaka-Oda A, Kenzo T, Kashimura S, Ninomiya I, Wang LH, Yoshikawa K, Fukuda K (2010) Physiological and morphological differences in the heterophylly of Sabina vulgaris Ant. in the semi-arid environment of Mu Us Desert, inner Mongolia, China. J Arid Environ 74:43–48

Winn AA (1999) The functional significance and fitness consequences of heterophylly. Int J Plant Sci 160:S113–S121

# Part IV
# The Life History Strategies of Plants Growing in Sand Shifting Environment

## Introduction

This chapter provides a detailed account of the life history of plants growing in the shifting sand environment of the Mu Us Sandy Land. It is necessary to discuss both how *Juniperus sabina*, a native natural vegetation species, adapts to the shifting sand environment through its strategies, and the growth and strategies of several species planted for the purpose of stabilizing shifting sands in this environment. In our series of studies that began in the late 1980s, *J. sabina* is explained in detail in Chap. 8 as a climax species of the Mu Us Sandy Land that forms the typical landscape of the fixed sand dunes. The latter plants are a group of species that have been planted, managed, and utilized around the periphery of shifting sand dunes for the purpose of conserving the local living and agricultural environments, and as a representative example, willows and poplars planted in and around the Mu Us Sandy Land are discussed in Chap. 9. Although various differences in water use physiology are observed between these two groups, commonalities can be found in terms of functional adaptations to drought and burial by shifting sand.

# Chapter 8
# Life History of *Juniperus sabina* L., a Climax Shrub Tree in the Mu Us Sandy Land, China

**Nobuhito Ohte**

## 8.1 *Juniperus sabina* in the Mu Us Sandy Land

*Juniperus* is an evergreen conifer. According to Adams and Schwarzbach (2013), there are about 75 existing species of the genus *Juniperus*, which are divided into three sections: section *Caryocedrus*, section *Juniperus*, and section *Sabina*. Section *Sabina* includes about 60 species, and its distribution is worldwide. According to Mao et al. (2010), the main part of this Section *Sabina* can be further divided into three clades: Serrate-leaf juniperus in North America, turbinate-seed cones, single-seeded, entire-leaf juniperus in eastern hemisphere, and multi-seeded, entire-leaf junipers. The distribution of this third clade is widespread, covering East Eurasia, Central Asia, Central Europe, North Africa, regions along the Great Rift Valley in East Africa, and the North American continent excluding eastern Canada, between 20° and 50° north latitude.

The *Juniperus sabina* L. discussed in this chapter belongs to the third clade. As shown in Chap. 5, the largest habitat of this species in sandy land is in the Mu Us Sandy Land in the central part of the Ordos Plateau (Wang et al. 2011). The native plant communities of the Mu Us Sandy Land are mainly composed of psammophytes, with the most dominant community being the *Artemisia ordosica* Krasch. community, which covers about 31% of the total area of the Mu Us Sandy Land. The vegetation of *A. ordosica* effectively prevents sand movement and provides favorable soil conditions for other perennial plant species. Therefore, in mature communities of *A. ordosica*, there are more than 90 accompanying species, and the communities of *J. sabina* can be seen as climax vegetation in sandy land, that is, a typical remnant of the natural landscape before land use changes.

Since the 1980s, numerous studies on the physiology and ecology of *J. sabina* have been conducted in this region, but the complete life history of this species had not been fully described or summarized before Ohte et al. (2021). Therefore, in this section, according to Ohte et al. (2021), this species' life history is explained based

 
K. Yoshikawa et al., *Dryland Vegetation in the Sand-Shifting Environment*,
https://doi.org/10.1007/978-981-95-4735-7_8

on the aforementioned previously published research and some new data obtained from recent field surveys. Following the overall purpose of this book, the description of the life history focuses on the adaptation strategies to this species' sand shifting environment, particularly in relation to the effects of quicksand, natural regeneration through seed reproduction, and the growth process of *J. sabina*. These findings are important in formulating ecological restoration plans against desertification in this region.

First, Fig. 8.1a, b illustrate the life history of this species over time, as revealed by previous research. These temporal changes were observed at sites shown in Fig. 8.2a, b. Interestingly, the life histories shown in Fig. 9.1a, b were observed at different locations shown in Fig. 8.2a, b, respectively.

Figure 8.1a Early growth, patch establishment, and expansion phases

I. The *S. cheilophila* community settles in the moist lowlands between the dunes.

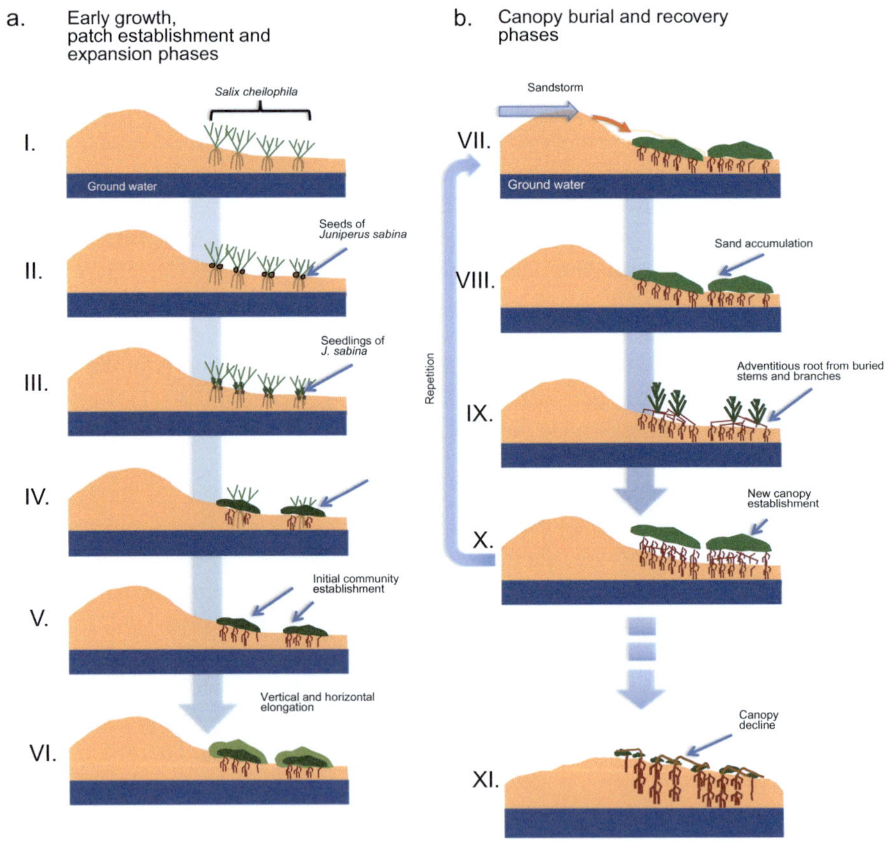


**Fig. 8.1** Life history and the sand dune environment. See the detailed explanation in the text

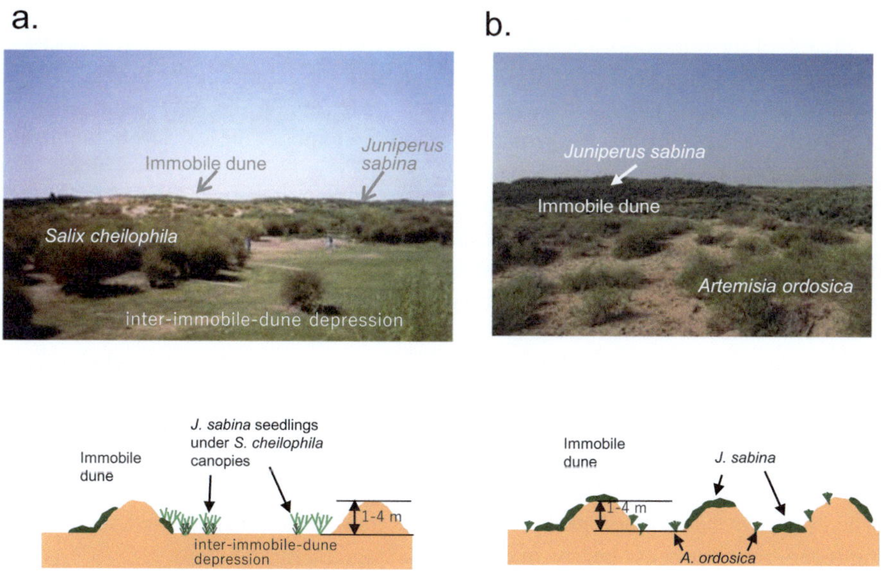


**Fig. 8.2** Typical landscape of *Juniperus sabina* habitat: (**a**) Early growth, patch establishment, and expansion phases and (**b**) canopy burial and recovery phases

II. The seeds of *J. sabina* are carried to the *S. cheilophila* community by birds and other animals. Strong seasonal winds and surface water flow during rainstorms may also potentially contribute to seed transport. However, the seeds are prepared for germination by the digestion of birds and animals.
III. Germination occurs, and the seedlings of *J. sabina* grow under the canopy of *S. cheilophila*.
IV. The patches of *J. sabina* grow vertically and horizontally, spreading from the canopy of *S. cheilophila*.
V. When the patches of *J. sabina* become much larger than the canopy of *S. cheilophila*, the individuals of *S. cheilophila* decline. Thus, the *J. sabina* community is established.
VI. The patches of *J. sabina* grow in both vertical and horizontal directions. Nutrient reproduction may occur to expand the patches horizontally.

Figure 8.1b Canopy burial and recovery phases

VII and VIII. The canopy of *J. sabina* is buried in quicksand during occasional large sandstorms.
IX. Adventitious roots grow from the buried trunk and branches. New shoots and branches grow on the new soil surface.
X. A new canopy is established on the new ground surface. Adventitious roots grow from the buried old trunk and creeping branches. A new root group is established in the soil layer near the ground surface.
XI. In mature *J. sabina* communities, some canopies may decline.

By repeating steps VII to X periodically, mature patches of *J. sabina* are formed on the top of the dunes. As a result, these dunes formed by this quicksand deposition process do not move with the wind like normal moving dunes. In mature *J. sabina* patches, no renewal occurs through seed reproduction, so the only mechanism to expand the canopy cover is the growth of creeping branches outward from the patch.

In some of the *J. sabina* patches where mature *J. sabina* patches have formed, it was found that the aboveground biomass was decreasing (XI). In these patches, the shading of the soil surface also decreased, and the light conditions under the canopy were improved. However, no germination of buried seeds or growth of seedlings was observed in these areas. This suggests that there may be factors inhibiting renewal by seed reproduction under their own canopy, such as soil moisture limitations due to parental tree transpiration and the aforementioned allelopathy, in addition to light conditions.

When the canopy declines and its covering effect is lost, the surface sand may start to move again. If this sand movement resumes on a large scale, lowlands between sand dunes are reformed by the movement of large sand dunes, and there is a possibility that *S. cheilophila* communities may become established. As a result, the regeneration process begins from the first stage mentioned above.

From the next section, I will describe what has been revealed from previous research for each phase of this life history.

## 8.2 Regeneration Process of *J. sabina*

### *8.2.1 Characteristic Seed and Nutrient Propagation Processes*

Many species of *Juniperus* can reproduce not only by seeds but also by nutrient propagation. Nutrient propagation of *Juniperus* has been studied in various species mainly for the purpose of artificial propagation (Berhe and Negash 1998; Negash 2002; Cope and Rupp 2013, etc.). It is known that *J. sabina* can reproduce nutritionally in natural ecosystems (Wang et al. 2011). *J. sabina* can produce adventitious roots where its creeping branches touch the ground. Parts where water and nutrients are sufficiently secured by adventitious roots can become independent from the original individual. There is no comprehensive study on whether seed propagation or nutrient propagation is dominant, but there are some case studies.

Wesche et al. (2005) and Ronnenberg (2005) showed that *J. sabina* in the desert steppe of poor soil regenerated mainly by nutrient propagation through creeping stems and adventitious roots. On the other hand, it was clear that the environmental conditions of *J. sabina* habitat in the Mu Us Sandy Land were unsuitable for seed germination, such as being poor in nutrients and having no shade from strong sunlight. An estimate of the age of a *J. sabina* individual suggested that the environment was suitable for seed propagation about 1800 years ago, and under current dry conditions, the survival of individuals/communities has been maintained by clonal propagation.

It has been shown that *J. sabina* in the Mu Us Sandy Land is also expanding the area of patches by nutrient propagation, but it has also been shown to have sufficient seed propagation ability. Genetic polymorphism of *J. sabina* populations in multiple locations in the Inner Mongolia Autonomous Region has been investigated. The genetic variation of the populations in the Mu Us Sandy Land was higher than that of other populations in the mountainous region (Zhang et al. 2005b; Lu et al. 2020). This suggests that while clonal propagation is frequent in the mountainous region, seed propagation also occurs in the Mu Us Sandy Land. Furthermore, Hong et al. (2006) investigated the degree of genetic polymorphism at different transition stages of *J. sabina* communities in the Mu Us Sandy Land, showing a high polymorphism rate in communities at the initial transition stage with many young individuals, indicating a strong influence of seed propagation. In communities at later stages where mature individuals are abundant, polymorphism is low, indicating an increased dependence on nutrient propagation.

As will be described later, there is a relationship between this transition stage and landscape, and it has been shown that the difference in the weight of seed propagation and nutrient propagation, as indicated by the degree of genetic polymorphism, is also reflected in the difference in habitats. That is, populations with more young individuals in the lowlands between dunes have stronger genetic differentiation than mature populations on fixed dune tops (Liang et al. 2007).

These things suggest that in the sand dune environment where sand moves, the renewal of community size by seed propagation occurs in specific landscapes and habitats.

### 8.2.2 Characteristics of Seeds and Seedlings

*J. sabina* is a dioecious woody shrub species, and the seeds are formed inside cones rich in fat. It is believed that birds living in the hairy crow sand habitat contribute to seed dispersal by eating the cones of *J. sabina* (Wang et al. 2005). The germination rate of these seeds is generally very low, both in the wild and under cultivation conditions (Tokuoka 1989; Zhang et al. 1997; Wesche et al. 2005). Tanaka-Oda et al. (2009) showed that the germination rate improves when the very hard seed coat tissue is soaked in sulfuric acid under appropriate temperature conditions (20–25 °C, Fig. 8.3). Winer (1983) and Chambers et al. (1999) showed that soaking in sulfuric acid has a similar effect to digestion in an animal's digestive system. Chambers et al. (1999) reported that the cones of the Juniperus genus are eaten by birds, sheep, goats, and wild rabbits.

Integrating these studies, it can be pointed out that the consumption of cones by animals and the digestion of the outer skin tissue may be one of the factors controlling the germination rate in the field. Buried seeds were found not only under the canopy of *J. sabina* but also under other shrubs in the vicinity, as described in the next paragraph. This fact supports the idea that seeds are being transported to places other than under the mature *J. sabina* canopy by birds or possibly other animals.

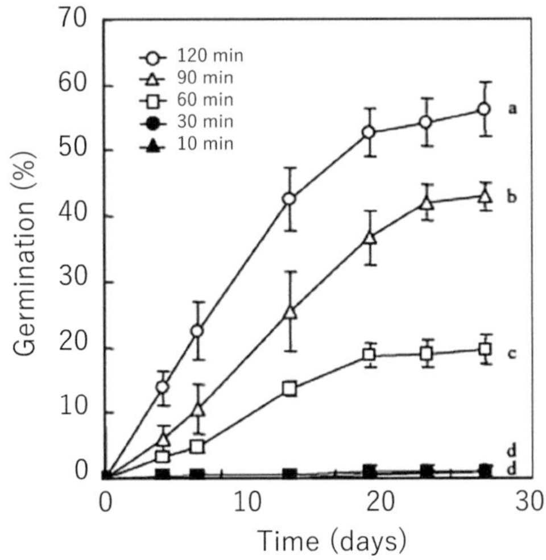


**Fig. 8.3** Germination rate of seeds soaked in sulfuric acid for different durations (mean ± SE). Different letters indicate significant differences ($P < 0.05$). (Modified from Tanaka-Oda et al. 2009)

The canopy of *J. sabina* is composed of creeping stems and leaves that directly cover the ground surface, so most of the fallen cones are right under the canopy. The distribution of buried seeds was investigated in the fixed sand dunes of the hairy crow sand habitat. Most of the surviving seeds were buried in the topsoil layer less than 4 cm under the canopy. A number of dead seeds, one to several orders of magnitude more than the living seeds, had accumulated in the same layer under the canopy (Fig. 8.4, Yamanaka et al. 1998). This suggests that the germination rate under field conditions was very low, which is corroborated by the extremely small number of seedlings growing under the canopy (Yamanaka et al. 1998; Wang et al. 2005).

In a survey at another site 16 km away from the same fixed sand dune, seedlings and buried seeds of *J. sabina* were observed under the canopy of specific woody shrub species *S. cheilophila*, *Salix psammophila*, and *J. sabina*. Young individuals (seedlings) were reported to be found only under the canopy of *S. cheilophila* and *S. psammophila* (Yamanaka et al. 2001; Wang et al. 2005; Ishii et al. 2006b). Although the environmental conditions are such that seed germination could occur, regeneration of *J. sabina* did not occur under these canopies.

In the vicinity of that location, *S. cheilophila* communities were growing mainly in the moist depressions between the fixed sand dunes, and their numbers were significantly higher than *S. psammophila* (Fig. 8.2a). Wang et al. (2005) reported that many *J. sabina* seeds were found under the canopy of *S. cheilophila* in the lower, wetter parts of the fixed sand dunes, and their seedlings grew vigorously in the same place. The immediate landscape of this location where *J. sabina* regeneration occurred had different characteristics from the landscape on the fixed sand dunes where mature *J. sabina* patches were established.

Although the seed density in the surface soil was much higher under *J. sabina* than under *S. cheilophila*, many buried seeds were found also under the *S. cheilophila* canopy. Cones were observed only under the *J. sabina* canopies (Ishii

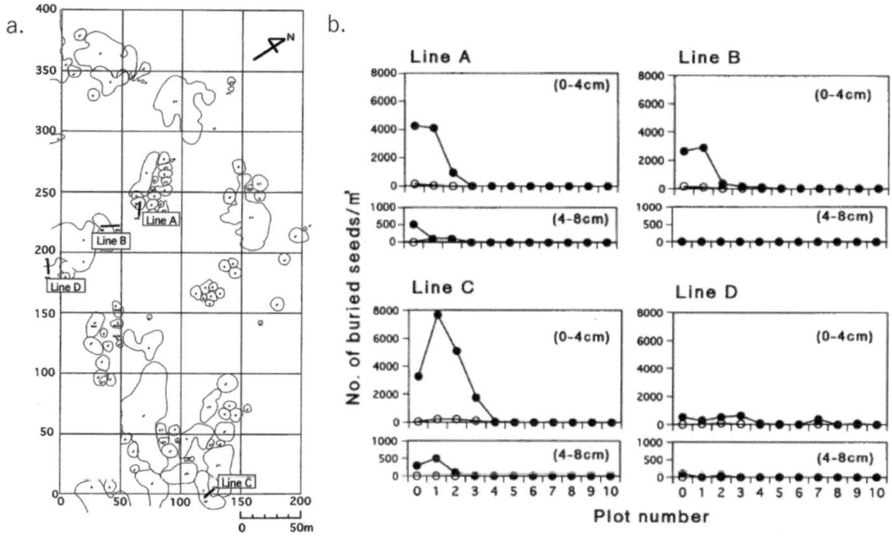


**Fig. 8.4** (**a**) Patch distributions of *Juniperus sabina* in a fixed sand dune plot. (**b**) Buried seed distribution on the transect lines. O: Survived buried seeds, ●: dead buried seeds. The transect lines (A to D) were set from points beneath the canopy to outside of each *Juniperus sabina* patch

personal communication). This suggested that the *J. sabina* seeds did not move to the *S. cheilophila* plants with their cone but were moved or transported as seeds. As mentioned in the previous section, birds and animals are considered potential carriers. *S. cheilophila* dominant stands were located in the lowest part of the fixed dune valleys, so the surface water flow that occurs during storms or strong winds might also carry seeds to the *S. cheilophila* stand area. Looking at the distribution of seeds buried under the canopy of *J. sabina* and *S. cheilophila*, the seeds were evenly distributed under the canopy of *J. sabina* but were concentrated near the trunk of *S. cheilophila* (Ishii, personal communication). On the other hand, seedlings were found under the canopy of both *J. sabina* and *S. cheilophila*, but seedling death was only found under the canopy of *J. sabina*, where no grown saplings were found. Seedlings clearly survived only under the canopy of *S. cheilophila* and were able to grow into saplings (Ishii 2006, Fig. 8.5a–d).

The reason why seedlings died under the canopy of *J. sabina* is that they could not get enough light resources (Yamanaka et al. 1998). Also, *J. sabina* adult individuals have a dense root system near the surface soil (Yoshikawa et al. 1988; Ohte et al. 2003). Adventitious roots from creeping branches play an important role in acquiring water from near the surface soil layer (Yang et al. 2014). This suggests that there may be competition for water between the parent individual and the seedlings under the canopy, and the seedlings may not be able to acquire enough water resources necessary for growth.

Another explanation is that the growth of seedlings was inhibited by allelopathy caused by *J. sabina*. Several other species of the genus *Juniperus*, including

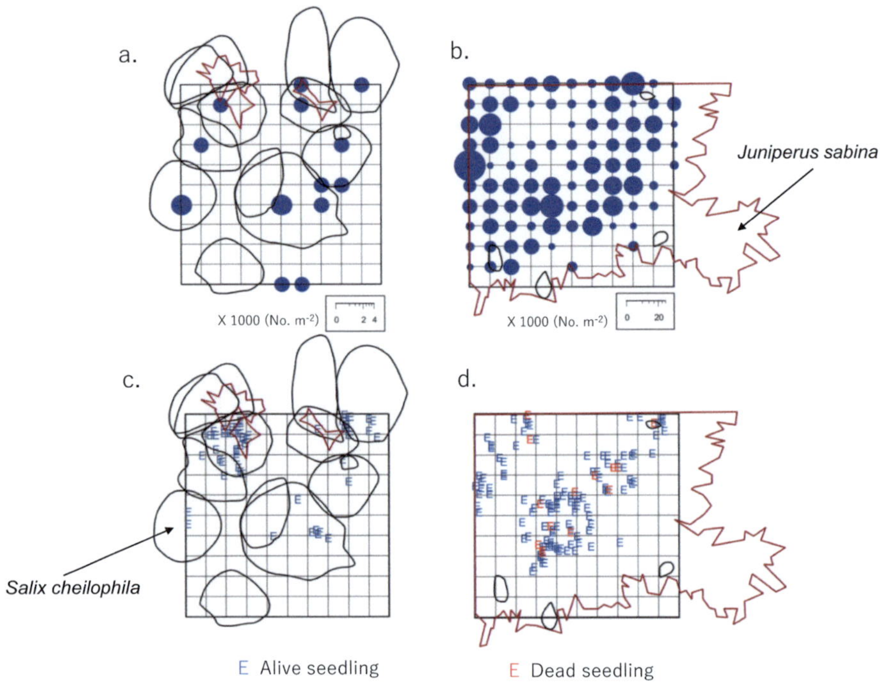


**Fig. 8.5** Seed distribution under canopies of (**a**) *Salix cheilophila* and (**b**) *Juniperus sabina*, and seedling distribution under canopies of (**c**) *Salix cheilophila* and (**d**) *Juniperus sabina*

*J. sabina*, show allelopathy (Jameson 1966; Lavin et al. 1968; Horman and Anderson 2003; Yager and Smeins 1999; Young and Bush 2009; Qin et al. 2021). Previous studies have not confirmed the fact that species of the genus *Juniperus* exert allelopathic activity and affect their own seedlings. The possibility of this self-inhibition by allelopathy is worth further investigation. These characteristics indicate that *J. sabina* cannot regenerate in the same place while the parent canopy is being maintained and grown. This reproductive pattern and growth process are deeply related to the movement of sand by wind, and the mechanism will be described in detail in later chapters.

### *8.2.3* **S. cheilophila** *as a Nurse Plant for* **J. sabina** *Seedlings*

*S. cheilophila* is a shrub species of the willow family. Each individual has multiple stems and does not have a main stem. The seeds have a tuft of cotton for effective wind dispersal. *S. cheilophila* requires relatively high soil moisture levels (Yuan and Li 2007), so its seeds were settled in the wet depressions between the dunes at the study site (Fig. 2c, Teraminami et al. 2013). In the case of *J. sabina* in this region, germination of seeds and growth of seedlings only occur under the canopy of *S. cheilophila*, hardly occur under the canopy of *S. psammophila*, and do not occur under other tree species (Wang et al. 2007). These observations suggest that *S. cheilophila* and *S. psammophila* act as nurse plants for the early growth of *J. sabina*. *S. cheilophila* Both the canopies of *J. sabina* reduce sunlight, preventing the conditions near the soil surface from becoming excessively hot and dry due to this shade (Fig. 8.6), and can create a microclimate favorable for seedling growth (Yamanaka et al. 2001; Ishii 2006). On the other hand, the light obtained under the canopy of *J. sabina* may be insufficient for photosynthesis by *J. sabina* seedlings (Yamanaka et al. 1998). Furthermore, from the same location as above, it was shown that the shading of *S. cheilophila* is effective in preventing photoinhibition damage caused by high light conditions during the cold winter period (Tanaka-Oda et al. 2010; Ishii et al. 2006b). This physiological characteristic is important for the survival of evergreen conifer seedlings in this region that experience a dry winter. Zhang (2006) investigated the mortality rate of *J. sabina* seedlings and concluded that low winter temperatures are the most effective factor in increasing mortality. Photoinhibition under low temperature winter conditions has been reported in various plant species (Demmig-Adams and Adams 1996; Gilmore 1997; Huner et al. 1998; Muller et al. 2001). These were mainly reported in semi-arid to arid environments. This resistance was an important characteristic for evergreen trees like *J. sabina* to adapt to dry and cold environments. The likelihood of successful regeneration increased as the nurse plant created shade during the seedling growth period, mitigating strong light.

This nursing effect was demonstrated by an experiment to remove the shade of *S. cheilophila* (Yamanaka et al. 2001). One year after tree removal, the survival rate

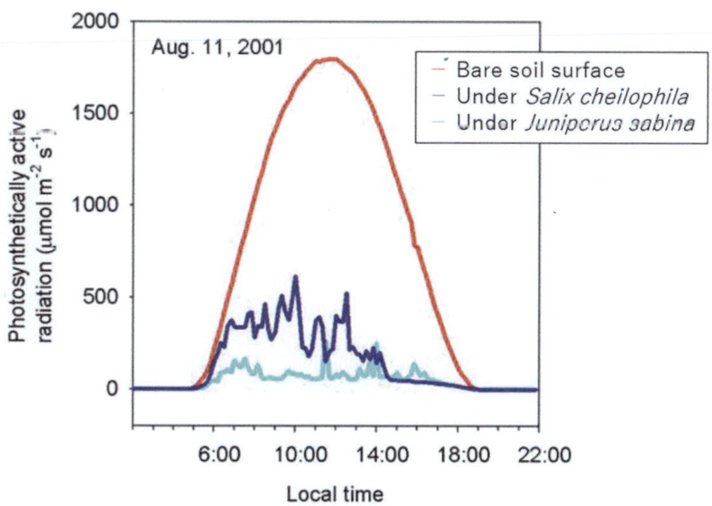


**Fig. 8.6** Photosynthetically active radiation under the canopy of *Salix cheilophila* and *Juniperus sabina*

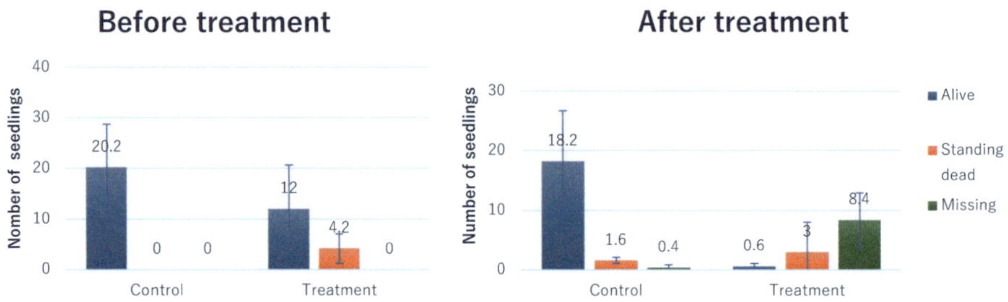


**Fig. 8.7** *Salix cheilophila* removing experiment. The results suggested that the young *Juniperus sabina* were stressed by high temperatures and extreme dryness due to the loss of canopy shade by *S. cheilophila*

of young *J. sabina* (height <50 cm) was 6%, but it was 88% in the sections where it was not removed. In the removal sections, 74% of the dead *J. sabina* individuals disappeared, and 26% died without falling. This suggests that the young plants were stressed by high temperatures and extreme dryness due to the loss of canopy shade by *S. cheilophila* (Fig. 8.7).

During the growth of young individuals, the canopy of *J. sabina* gradually expands from the edge of the canopy of *S. cheilophila*, resulting in part of the canopy being exposed to strong sunlight, high temperatures, and low humidity. As it grows from a seedling to a young shrub, *J. sabina* develops resistance to drought conditions. The water use efficiency evaluated by the leaf tissue's $\delta^{13}C$ increased monotonically with the expansion of the canopy (Figs. 8.8 and 8.9, Ohte et al. 2021). The $\delta^{13}C$ value of the largest individual sampled in this survey was similar to the previous report of adult individuals (Ohte et al. 2003). At the same time, *J. sabina* changes the shape of its leaves from needle-like to scale-like. This morphological change minimizes transpiration from the cuticle, thereby enhancing drought resistance (Tanaka-Oda et al. 2010). Furthermore, *J. sabina* also develops resistance to photoinhibition under the strong sunlight conditions present at this stage. To prevent damage from photoinhibition, *J. sabina* changes the composition of photosynthetic pigments in its leaves (Ishii et al. 2006a, b).

Thus, in the lowlands between the dunes in the fixed dunes, the initial stages of community formation of *J. sabina* where seeds germinate and seedlings grow into young plants were observed. On the other hand, in the fixed dunes at another location, there were many patches of mature individuals. Despite no difference in site conditions between these two places, no *S. cheilophila* communities were found at the latter location. This is because *J. sabina* functioned as a nurse plant in the early stages of growth of *S. cheilophila* communities eventually declined, suggesting that *J. sabina* communities have become established. One possible explanation for this transition is that plants may be competing for water resources. In the next chapter, we will explain that *J. sabina* can utilize soil moisture not only in the surface layer but also in the deep layer close to the groundwater level.

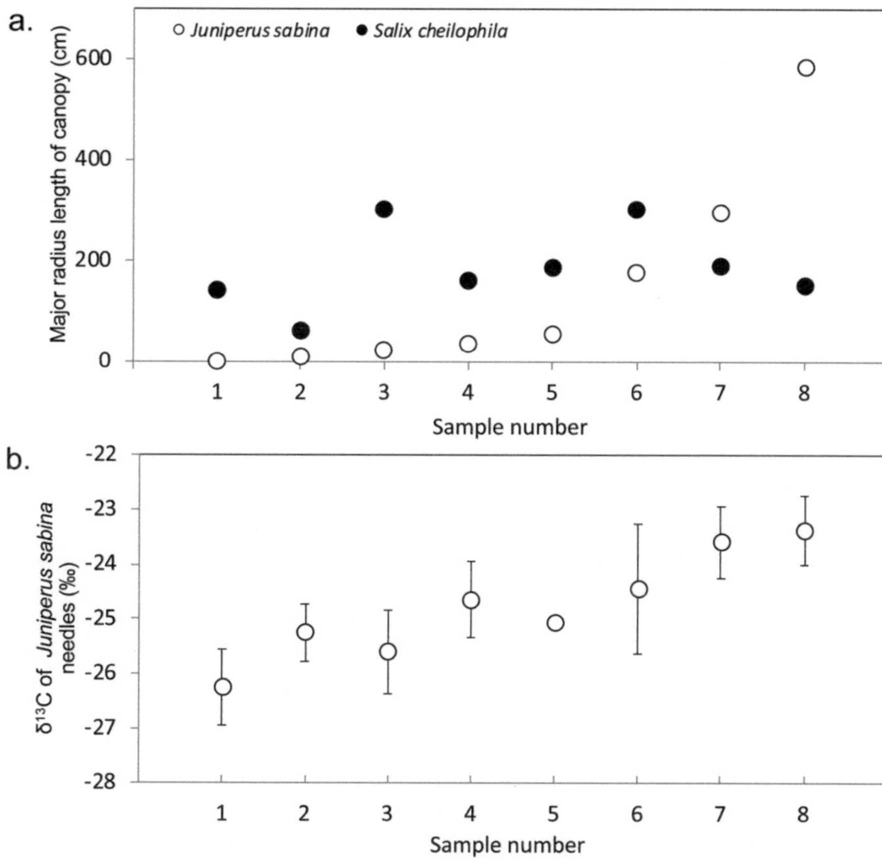


**Fig. 8.8** (**a**) Changes in canopy size during seedling growth of *Juniperus sabina* and *Salix cheilophila* and (**b**) the $\delta^{13}C$ values of *J. sabina* needles (mean ± standard deviation). For the samples 7 and 8, the part of the *J. sabina* canopy expanded out from the edges of the canopies of *S. cheilophila*. The survey was conducted in August 2001

## 8.3 Growth Characteristics of Adult *J. Sabina* Patches

### *8.3.1 Growth Rate and Environmental Conditions*

Miki et al. (2016) observed the horizontal expansion of creeping branches of adult *J. sabina* patches at this location. As a result, it was shown that the annual variation in growth rate strongly correlates with the variation in annual precipitation, suggesting a strong regulation of water availability on the growth rate of this species (See Fig. 6.7 in Chap. 6). The average growth rate (horizontal expansion of creeping branches) from 1997 to 2003 was 13 cm/year. Based on this observation, a simple estimate suggests that it would take about 40 years to form a patch with a diameter of about 10 m. This growth rate estimate was roughly confirmed by direct age

**Fig. 8.9** Seedlings, young and mature patch of *Juniperus sabina* (Photographed by N. Ohte). The pictures (**a**–**d**) correspond to the sample numbers 1, 3, 6, and 8, respectively

determination using $^{14}$C dating of wood samples from adult patches at the same location (Ohte et al. 2021, Table 1). In this measurement, wood samples were taken from the thickest central trunk of three adult patches using a growth cone. The average estimated age of the adult communities with a diameter of about 10 m was 36.8 ± 13.2 years.

### *8.3.2 Physiological and Morphological Characteristics for Drought Tolerance*

The water use characteristics and strategies of adult *J. sabina* have been intensively investigated and discussed by Kobayashi (1992), Li et al. (2003), He and Zhang (2003), Ohte et al. (2003), Li et al. (2004), Yang et al. (2008), Wang et al. (2011), Murakami et al. (2016), and Miki et al. (2016, 2018). Ohte et al. (2003) showed that *J. sabina* has a unique high water use efficiency among woody species growing in the same site. Furthermore, the root system of *J. sabina* is evenly distributed from the surface to near the groundwater level (about 100 cm deep), and it was shown that it can absorb soil water from various depths of the soil profile (Ohte et al. 2003).

This vertical root distribution was observed not only in the center of the *J. sabina* patch but also in the periphery structured by the growth of adventitious roots from the ground-contacting parts of creeping branches grown in the periphery. Yang et al. (2014) and Miki et al. (2016, 2018) provide detailed descriptions of the use of adventitious roots for soil water use by *J. sabina*. Adventitious roots originating from creeping branches usually absorb soil water from the surface soil. When the period without precipitation is long and the moisture in the surface soil is limited, the adventitious roots supply water transported from the main roots in the patch center to the surface soil at night (Yang et al. 2014; Miki et al. 2016, 2018). On the other hand, the main roots can always access water in deeper layers. This phenomenon is called hydraulic redistribution (e.g., Richards and Caldwell 1987; Caldwell and Richards 1989). *J. sabina* can maintain the drought tolerance of the entire patch through this root function. Also, Li et al. (2003) demonstrated that drought stress inhibits shoot elongation of *J. sabina* and promotes vertical growth of the root system. As a result, it was found that as soil moisture content decreases, a strategy is adopted to widen the range of soil for securing water.

As mentioned above, the ability of these creeping branches to form adventitious roots where they contact the ground increases the opportunity to obtain soil water. Also, if the supply of water and nutrients from the main root is cut off, this function allows the creeping stem to survive as an independent individual (Tokuoka 1989). In other words, the function of drought tolerance through adventitious root formation serves as the basis for vegetative reproduction. Indeed, when patch expansion as mentioned in the previous section occurs, vegetative reproduction occasionally takes place, and it is thought that the formation of large colonies progresses.

As such, *J. sabina* has other important strategies to withstand the severe restrictions of soil moisture, due to both physiological and morphological adaptations. Mature patches of *J. sabina* usually have a very dense canopy (Yoshikawa et al. 1988), limiting evaporation from the soil, and the availability of soil moisture is higher under the canopy cover than outside (Yamanaka et al. 1999).

Furthermore, this dense canopy accelerates nutrient accumulation under the canopy by preventing leaves and fine sand (silt and cray) from being blown away by the wind. Hirobe et al. (2001) reported that the accumulation of organic matter derived from litter is about 6 times higher under the canopy than outside and detailed the physicochemical properties of these "fertile islands."

Many reports to date have shown that ground cover by the canopy not only accumulates nutrients by preventing leaf scattering but also aids in the survival of seeds and seedlings (Buckley 1996; Singh and Rathod 2002; Wiggs et al. 1995). In contrast, as explained earlier, seedlings of *J. sabina* cannot continue to grow under the parent's canopy due to resource limitations, such as light.

### 8.3.3 Canopy Growth and Shifting Sand

As mentioned earlier, fixed sand dunes covered by *J. sabina* patches have undulations of about 1 to 4 m in relative height, and many growing *J. sabina* patches spread to the top of the dune as if crawling up the slope. To clarify the mechanism behind this phenomenon, we investigated the spatial distribution of roots under *J. sabina* patches at the top of the dune (Fig. 8.10, Ohte et al. 2021). Under the center of this patch, there were four distinct soil layers where roots were concentrated. All roots within these layers were alive.

These layered root groups are presumed to have once existed just below the soil surface, directly capturing soil water supplied by precipitation. When moving sand covers the patch, *J. sabina* generates adventitious roots from stems and branches that were above ground but are now buried, constructing a root layer just below the new ground surface. Then, the stem extends to the soil surface and forms a canopy above the new ground. A few years later, another sand flow event occurs, and the canopy is completely covered with sand again. If such events are repeated, the canopy that has received sand deposition always ends up at the top of the slope, and multiple distinct root layers are formed. The soil particle size in the root layer was

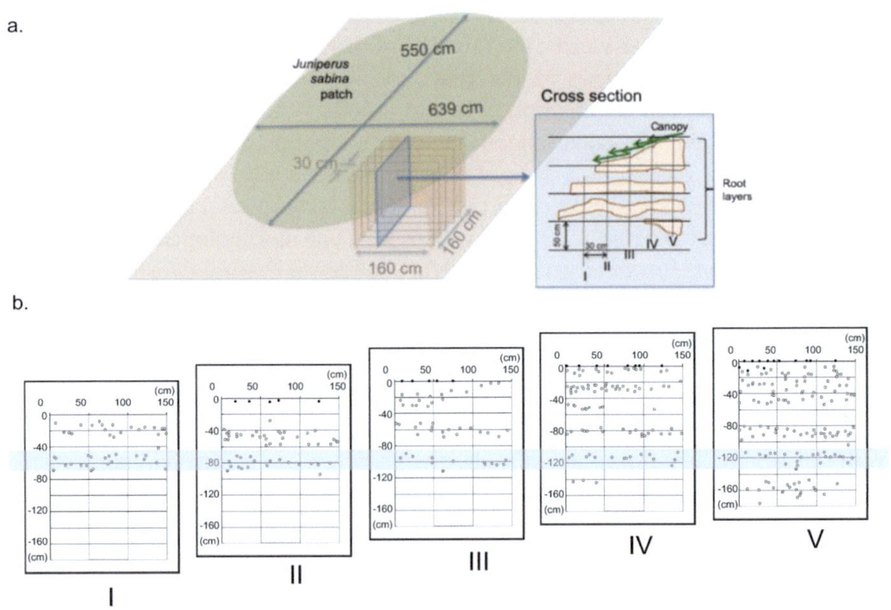


**Fig. 8.10** (**a**) Trench was established for surveying the root distribution under a *J. sabina* patch at an immobile dune. Five cross-sectional soil profiles were shaped for the observation of roots distribution. (**b**) Roots layer distributions in the soil profiles. The gray hatch indicates the same vertical level in the soil profile. Open circles indicate the positions of roots and closed circles around the soil surface indicate the positions of prostrate branches. The survey was conducted in August 2001

smaller than the soil between the root layers. These fine soil layers indicated that fine sand was captured and deposited within the ground covered by the dense *J. sabina* canopy. This is further evidence supporting Hirobe et al.'s (2001) discussion on soil accumulation when a *J. sabina* canopy is established.

The vertically spreading root system mentioned in the previous section can be explained by the above dynamics of *J. sabina* and sand. That is, mature patches of *J. sabina* on the dunes in the fixed sand dune area experience burial by sand flow several times over decades. Each time, *J. sabina* extends its trunk and branches upward, forming a new canopy for primary production on the new ground surface. At the same time, adventitious roots actively grow from the buried trunk and branches to obtain water from the soil layer near the ground surface. The experience of burial by sand deposition enhances the availability of soil moisture throughout the patch, promoting growth. This was demonstrated by Miki et al. (2016), indicating that the annual rate of sand deposition is an important control factor for the annual growth rate of fixed dune patch size. Furthermore, the undulation of fixed dunes covered by the mature *J. sabina* canopy suggests the possibility of sand accumulation and repeated canopy formation in *J. sabina* communities.

The ratio of aboveground to belowground biomass in trees is widely recognized to be within a certain range for each species and habitat, depending on the individual's strategy to balance assimilative and non-assimilative parts to maintain the tree body.

In the case of *J. sabina*, it is tentatively considered that root death may occur in response to the reduction of the assimilative part of the canopy when the canopy is buried in sand and the aboveground biomass decreases. No direct investigation has been conducted on the amount and rate of root death in *J. sabina* in this region. However, Li et al. (2003) conducted growth experiments with *J. sabina* under several different growth conditions and showed that the T/R ratio of *J. sabina* varies greatly depending on soil moisture conditions. The spatial distribution of the root system shown in Fig. 8.10 indicates that the older, lower root system is also used for moisture acquisition. This suggests that it is reasonable to assume that soil water is present in the deep soil (Ohte et al. 2003; Miki et al. 2016). Furthermore, it can be concluded that a key characteristic of this species is its ability to survive even when the T/R ratio temporarily decreases. This characteristic ensures long-term moisture acquisition, so it is thought that once the canopy biomass decreases, it can recover and grow further. The observation results shown by Miki et al. (2016) that sand coverage promotes canopy growth support this assumption.

Based on these observations, it can be concluded that three main strategies are taken to obtain the water necessary for patch growth from a wide range of soil layers: (1) suppression of evaporation by dense canopy, (2) active development of adventitious roots from creeping branches and hydraulic redistribution through these roots, and (3) active development of water-absorbing root system using adventitious roots from stems and branches buried in the ground when quicksand covers the canopy. Furthermore, it can be pointed out that a close nutrient cycle occurs between the canopy and the topsoil due to the ground cover provided by the dense canopy. The impact of soil water and nutrient availability on leaf growth and shape has been investigated by He and Zhang (2003), showing that soil water availability is the most important factor controlling leaf growth.

## 8.4 Implications of the Life History of *J. Sabina* Adapted to Quicksand Environments

As mentioned so far, the processes of regeneration, growth, and community establishment based on seed reproduction are thought to be closely related to the movement of sand by wind. The most important characteristic that allows *J. sabina* to adapt to this moving sand environment is its high plasticity in maintaining a shrub form that can quickly generate adventitious roots from parts buried in deposited sand.

This pattern of growth and regeneration suggests that in the development process of *J. sabina* communities, moving sand dunes continuously existed around the community. The observation of layered root systems (Fig. 8.10) indicates that during the development process of the community, the movement of sand by wind was continuous and sometimes very active. The large-scale movement of sand that buries the canopy suggests that it has occurred several times in less than 40 years, which is the estimated patch age of the study subjects.

Most of the research on plants adapted to growth and regeneration in the environment of shifting sand dunes focuses on grasses of the Poaceae family growing in coastal environments (Maun 1998; Musila et al. 2001; Dech and Maun 2006). For tree species, there is research on the regeneration of willows in coastal dunes on the west coast of the UK (Ranwell 1960). In inland areas, long-term monitoring of the recovery of degraded vegetation in the Horqin Sandy Land has been conducted (Zhang et al. 2005a). These previous studies emphasize the importance of adaptation or resistance to being buried in sand as a functional characteristic of the dominant plants in each location. The generation of adventitious roots from organs that were once above ground is an important mechanism of this adaptation. Due to the high plasticity of above-ground organs, plants can acquire enough water to survive in semi-arid sandy lands.

*J. sabina* has been shown to have acquired this characteristic in the Mu Us Sandy Land. The unique reproductive behavior of *J. sabina* through seed reproduction is thought to be a result of adaptation to the environment of shifting sand. In an environment where leaves and sand are easily moved by seasonal strong winds, *J. sabina* can form a dense canopy near the ground and develop adventitious roots to prevent the loss of sand and leaves and secure nutrients and water resources. This adaptation allows adult patches to maintain above-ground biomass even when the canopy is buried in shifting sand.

However, these dense canopies do not allow enough light to penetrate for seedlings to grow, so reproduction is only possible where *S. cheilophila* or *S. psammophila* grows. These characteristics may be a trade-off with *J. sabina* having acquired the functions necessary for long-term survival in maturity. In other words, it may have lost the ability to raise seedlings under its own canopy. As a result,

*J. sabina* can reproduce by seeds in places separate from the mother community. As mentioned earlier, vegetative reproduction by the elongation of nutrient branches and the rooting of adventitious roots was thought to be effective in maintaining and spatially expanding the community in the place where the mother community exists. In another variant of *J. sabina* described by Wesche et al. (2005) and Ronnenberg (2005), only the function of vegetative reproduction remains, enabling the long-term maintenance of the community.

Reproduction is necessary in places different from the mother community, but the shifting sand environment of the Mu Us Sandy Land left environmental conditions that allow seed reproduction. Despite semi-arid climatic conditions, the formation of another tree community that can serve as a nurse plant for *J. sabina* seedlings and the abundance of groundwater and soil moisture in the lowlands between sand dunes made this seed reproduction possible.

So far, there have been several macro studies indicating that desertification has been progressing since before the 1980s (e.g., Rost et al. 2003; Karnieli et al. 2014), but according to major literature, the details of the landscape of the Mu Us Sandy Land before the 1980s were unknown. However, even now, there are scattered communities of *J. sabina* consisting of mature patches and young patches including seedlings in various places in the Mu Us Sandy Land. Furthermore, it was found that there are individuals of *J. sabina* that are over 50 years old in some regions (see Ohte et al. 2021, Table 1). These suggest that landscapes such as lowlands and wetlands between sand dunes, which are necessary for seed reproduction, have continuously existed since before the 1980s.

## References


Adams RP, Schwarzb AE (2013) Phylogeny of Juniperus using nrDNA and four cpDNA regions. Phytrogia 95(2):179–187

Berhe D, Negash L (1998) Asexual propagation of Juniperus procera from Ethiopia: a contribution to the conservation of African pencil cedar. For Ecol Manag 112(1):179–190. https://doi.org/10.1016/S0378-1127(98)00327-2

Buckley RC (1996) Effects of vegetation on the transport of dune sand. Ann Arid Zone 35(3):215–223. https://www.researchgate.net/publication/29470148_Effects_of_Vegetation_on_the_Transport_of_Dune_Sand

Caldwell MM, Richards JH (1989) Hydraulic lift: water efflux from upper roots improves effectiveness of water uptake by deep roots [journal article]. Oecologia 79(1):1–5. https://doi.org/10.1007/bf00378231

Chambers J, Vander Wall S, Schupp E (1999) Seed and seedling ecology of piñon and juniper species in the pygmy woodlands of western North America. Bot Rev 65(1):1–38. https://doi.org/10.1007/BF02856556

Cope KR, Rupp LA (2013) Vegetative propagation of Juniperus osteosperma (Utah Juniper) by cuttings. Nat Plants J 14(2):76–84. https://doi.org/10.3368/npj.14.2.76

Dech JP, Maun MA (2006) Adventitious root production and plastic resource allocation to biomass determine burial tolerance in Woody plants from Central Canadian coastal dunes. Ann Bot 98(5):1095–1105. https://doi.org/10.1093/aob/mcl196

Demmig-Adams B, Adams WW III (1996) The role of xanthophyll cycle carotenoids in the protection of photosynthesis. Trends Plant Sci 1(1):21–26. https://doi.org/10.1016/S1360-1385(96)80019-7

Gilmore AM (1997) Mechanistic aspects of xanthophyll cycle-dependent photoprotection in higher plant chloroplasts and leaves. Physiol Plant 99(1):197–209. https://doi.org/10.1111/j.1399-3054.1997.tb03449.x

He W-M, Zhang X-S (2003) Responses of an evergreen shrub Sabina vulgaris to soil water and nutrient shortages in the semi-arid Mu Us Sandland in China. J Arid Environ 53(3):307–316. https://doi.org/10.1006/jare.2002.1051

Hirobe M, Ohte N, Karasawa N, Zhang G-S, Wang L-H, Yoshikawa K (2001) Plant species effect on the spatial patterns of soil properties in the Mu-Us desert ecosystem, Inner Mongolia, China. Plant Soil 234(2):195–205. https://doi.org/10.1023/A:1017943030924

Hong Y, Wang L, Guosheng Z, Bayaer E, Xiaorong L (2006) Genetic diversity of Sabina vulgaris populations at different succession stages. Chin J Appl Ecol 17(11):2006–2010

Horman CS, Anderson VJ (2003) Understory species response to Utah Juniper litter. J Range Manag 56(1):68–71. https://doi.org/10.2307/4003884

Huner NPA, Öquist G, Sarhan F (1998) Energy balance and acclimation to light and cold. Trends Plant Sci 3(6):224–230. https://doi.org/10.1016/S1360-1385(98)01248-5

Ishii Y (2006) Study oncommunity dynamics and light torelance of Sabina vulgaris Ant. In: The Mu-Us sandy land, Inner Mongolia, China. Okayama University

Ishii Y, Sakamoto K, Wang L-h, Yoshikawa K (2006a) Seasonal changes in pigment composition of Sabina vulgaris Needles in the Mu Us Sandy Land of northern China. J Arid Land Stud 16(2):85–93

Ishii Y, Sakamoto K, Yamanaka N, Wang L, Yoshikawa K (2006b) Light acclimation of needle pigment composition in Sabina vulgaris seedlings under nurse plant canopy. J Arid Environ 67(3):403–415. https://doi.org/10.1016/j.jaridenv.2006.03.004

Jameson DA (1966) Pinyon-Juniper litter reduces growth of Blue Grama. J Range Manag 19(4):214–217. https://doi.org/10.2307/3895650

Karnieli A, Qin Z, Wu B, Panov N, Yan F (2014) Spatio-Temporal Dynamics of Land-Use and Land-Cover in the Mu Us Sandy Land, China, Using the Change Vector Analysis Technique. Remote Sensing 6:9316–9339. https://www.mdpi.com/2072-4292/6/10/9316

Kobayashi T (1992) Study on ecophysiological properties of afforestation plants in the Mu-Us desert in China. Kyoto University

Lavin F, Jameson DA, Gomm FB (1968) Juniper extract and deficient aeration effects on germination of six range species. J Range Manag 21(4):262–263. https://doi.org/10.2307/3895828

Li Y, Hayashi M, Sakamoto K, Yoshikawa K (2003) The effect of water stress on root growth of Sabina vulgaris Ant. and Juniperus chinenses ver. kaizuka Hort. J Jpn Soc Reveget Technol 29(1):107–112 (in Japanese with English summary)

Li Y, Hayashi M, Miki N, Sakamoto K, Yoshikawa K (2004) The effect of water stress on transpiration and water uptake of Sabina vulgaris Ant. and Juniperus chinenses ver. kaizuka Hort. J Jpn Soc Reveget Technol 30(1):98–103 (in Japanese with English summary)

Liang X, Zhang G, Bao B, Wang L, Hong Y (2007) Genetic diversity of Sabina vulgaris under different microhabitat in Maowusu Sandyland. J Inner Mongolia Agric Univ 28(2):67–72

Lu D, Zhang G, Li Y, Guo W, Zhang Y, Yu L, Hu S, Ao M, Wang A (2020) Genetic diversity and evolutionary history analysis of natural populations of Juniperus sabina L. Plant Sci J 38(2):151–161 (in Chinese with English summary). https://doi.org/10.11913/PSJ.2095-0837.2020.20151

Mao K, Hao G, Liu J, Adams RP, Milne RI (2010) Diversification and biogeography of Juniperus (Cupressaceae): variable diversification rates and multiple intercontinental dispersals. New Phytol 188:254–272. https://doi.org/10.1111/j.1469-8137.2010.03351.x

Maun MA (1998) Adaptations of plants to burial in coastal sand dunes. Can J Bot 76(5):713–738. https://doi.org/10.1139/b98-058

Miki NH, Kubori F, Yang L, Matsuo N, Zhang G, Wang L, Yoshikawa K (2016) Effective water use and growth of a prostrate lifeform shrub, Juniperus Sabina, in semiarid areas of China. J Arid Land Stud 26(3):95–99. https://doi.org/10.14976/jals.26.3_95

Miki NH, Kubori F, Yang L, Matsuo N, Zhang G, Wang L, Yoshikawa K (2016) Effective water use and growth of a prostrate lifeform shrub, Juniperus sabina, in semiarid areas of China. J Arid Land Stud 26(3):95–99

Miki NH, Sato K, Aoki M, Yang L, Matsuo N, Zhang G, Wang L, Yoshikawa K (2018) Water movement via adventitious roots of the prostrate shrub Juniperus sabina in semiarid areas of China. Acta Hortic 1222:137–146. https://doi.org/10.17660/ActaHortic.2018.1222.19

Müller P, Li X-P, Niyogi KK (2001) Non-photochemical quenching. A response to excess light energy. Plant Physiol 125(4):1558–1566. https://doi.org/10.1104/pp.125.4.1558

Murakami Y, Miki NH, Yang L, Zhang G, Wang LH, Yoshikawa K (2016) Water transport properties of seven woody species from the semi-arid mu us Sandy land, China [journal article]. Landsc Ecol Eng 12(2):209–220. https://doi.org/10.1007/s11355-015-0290-2

Musila WM, Kinyamario JI, Jungerius PD (2001) Vegetation dynamics of coastal sand dunes near Malindi, Kenya. Afr J Ecol 39(2):170–177. https://doi.org/10.1046/j.1365-2028.2001.00294.x

Negash L (2002) Successful vegetative propagation techniques for the threatened African pencil cedar (Juniperus procera Hoechst. ex Endl.). For Ecol Manag 161(1):53–64. https://doi.org/10.1016/S0378-1127(01)00501-1

Ohte N, Koba K, Yoshikawa K, Sugimoto A, Matsuo N, Kabeya N, Wang L (2003) Water utilization of natural and planted trees in the semiarid desert of Inner Mongolia, China. Ecol Appl 13(2):337–351. https://doi.org/10.1890/1051-0761(2003)013[0337:WUONAP]2.0.CO;2

Ohte N, Miki NH, Matsuo N, Yang L, Hirobe M, Yamanaka N, Ishii Y, Tanaka-Oda A, Shimizu M, Zhang G, Sakamoto K, Wang L, Yoshikawa K (2021) Life history of *Juniperus sabina* L. adapted to the sand shifting environment in the Mu Us Sandy Land, China: a review. Landsc Ecol Eng 17(3):281–294. https://doi.org/10.1007/s11355-021-00453-y

Qin L, Miki NH, Zhang G, Fujii Y (2021) Allelopathic effects of the revegetation species Juniperus sabina L. in semiarid areas of China. Landsc Ecol Eng 17(3):245–251. https://doi.org/10.1007/s11355-019-00395-6

Ranwell D (1960) Newborough Warren, Anglesey: II. Plant Associes and succession cycles of the sand dune and dune slack vegetation. J Ecol 48(1):117–141. https://doi.org/10.2307/2257311

Richards JH, Caldwell MM (1987) Hydraulic lift: substantial nocturnal water transport between soil layers by Artemisia tridentata roots [journal article]. Oecologia 73(4):486–489. https://doi.org/10.1007/bf00379405

Ronnenberg K (2005) Reproductive ecology of two common woody species, Juniperus sabina and Artemisia santolinifolia, in mountain steppes of southern Mongolia. In Erforschung biologischer Ressourcen der Mongolei, Proceedings of the symposium "Ecosystem Research in the Arid Environments of Central Asia: Results, Challenges, and Perspectives," Ulaanbaatar, Mongolia, June 23–24, 2004. (Vol. 9, pp. 207–223). Institut für Biologie der Martin-Luther-Universität Halle-Wittenberg

Rost KT, Böhner J, Pörtge K-H (2003) Landscape degradation and desertification in the Mu Us Shamo, Inner Mongolia—an ecological and climatic problem since historical times? Erdkunde 57(H.2):110–125. http://www.jstor.org/stable/25647546

Singh G, Rathod TR (2002) Plant growth, biomass production and soil water dynamics in a shifting dune of Indian desert. For Ecol Manage 171(3):309–320. https://doi.org/10.1016/S0378-1127(01)00791-5

Tanaka-Oda A, Kenzo T, Fukuda K (2009) Optimal germination condition by sulfuric acid pretreatment to improve seed germination of Sabina vulgaris Ant. J For Res 14(4):251–256. https://doi.org/10.1007/s10310-009-0129-5

Tanaka-Oda A, Kenzo T, Kashimura S, Ninomiya I, Wang LH, Yoshikawa K, Fukuda K (2010) Physiological and morphological differences in the heterophylly of Sabina vulgaris ant. In the semi-arid environment of Mu Us Desert, Inner Mongolia, China. J Arid Environ 74(1):43–48. https://doi.org/10.1016/j.jaridenv.2009.07.013

Teraminami T, Nakashima A, Ominami M, Yamamoto M, Zhang G, Yoshikawa K (2013) Effects of sand burial depth on the root system of Salix cheilophila seedlings in mu us Sandy land, Inner Mongolia, China. Landsc Ecol Eng 9(2):249–257. https://doi.org/10.1007/s11355-012-0205-4

Tokuoka M (1989) On the propagation of Cha Zi Yuan Bai (Sabina vulgaris) distributed in the Mao Wu Su sands of inner Mongolia, China. Bull Kyoto Prefect Univ For 33:7–11 (in Japanese with English summary)

Wang Z, Zhang GS, Wang LH, Hao YL, Wen GS (2005) Seed yield, seed bank and regeneration of natural Sabina vulgaris is community in Mu Us sandland. J Arid Land Resour Environ 19(3):195–200 (in Chinese with English summary)

Wang Z, Zhang GS, Wang LH, Hao YL, Wang M (2007) Microhabitat characteristics of natural regeneration of Sabina vulgaris community in Mu Us Sandland. J Arid Land Resour Environ 21(4):156–163

Wang, L., Zhang, G., Wen, G., Li, Y., & Yoshikawa, K. (2011). Physiological and ecological characteristics of Juniperus sabina: commmunity restoration and establishment (in Chinese). Scientific Press

Wesche K, Ronnenberg K, Hensen I (2005) Lack of sexual reproduction within mountain steppe populations of the clonal shrub Juniperus sabina L. in semi-arid southern Mongolia. J Arid Environ 63(2):390–405. https://doi.org/10.1016/j.jaridenv.2005.03.014

Wiggs GFS, Thomas DSG, Bullard JE, Livingstone I (1995) Dune mobility and vegetation cover in the Southwest Kalahari desert. Earth Surf Process Landf 20(6):515–529. https://doi.org/10.1002/esp.3290200604

Winer N (1983) Germination of pretreated seed of mesquite (Prosopis chilensis) under arid conditions in northern Sudan. For Ecol Manag 5(4):307–312. https://doi.org/10.1016/0378-1127(83)90035-X

Yager LY, Smeins FE (1999) Ashe Juniper (Juniperus Ashei: Cupressaceae) canopy and litter effects on understory vegetation in a Juniper-oak savanna. Southwest Nat 44(1):6–16. http://www.jstor.org/stable/30055397

Yamanaka N, Wang LH, Yoshikawa K (1998) Seed production and seed bank of Sabina vulgaris Ant. in the Mu Us Desert, Inner Mongolia, China. J Jpn Soc Reveget Technol 23(4):211–219 (in Japanese with English summery). https://doi.org/10.7211/jjsrt.23.211

Yamanaka N, Wang LH, Yoshikawa K (1999) Microenvironment of safe site for Sabina vulgaris Ant. Seedlings in the Mu-us Desert, Inner Mongolia, China. J Jpn Soc Reveget Technol 25(4):437–442 (in Japanese with English). https://doi.org/10.7211/jjsrt.25.437

Yamanaka N, Zhang GS, Sakamoto K, Wang LH, Yoshikawa K (2001) Effect of nurse plant removal on the seedling survival of Sabina vulgaris Ant. in the MU US desert, Inner Mongolia, China. J Jpn Soc Reveget Technol 27(1):120–124 (in Japanese with English summary). https://doi.org/10.7211/jjsrt.27.120

Yang L, Miki N, Harada N, Hirobe M, Sakamoto K, Yoshikawa K (2008) Response of water-use properties and growth of *Sabina vulgaris* Ant. in different water conditions. J Environ Sci Sustain Soc 2:27–34

Yang L, Miki NH, Matsuo N, Zhang G, Wang L, Yoshikawa K (2014) Contribution of adventitious roots to water use strategy of Juniperus sabina in a semiarid area of China. J Agri Technol A 4:251–259

Yoshikawa K, Shu-qin L, Kobashi S, Masuda T (1988) A preliminary survey of plant biomass at the Mu Us Desert in Inner Mongolia. Bull Reveget Res 10:66–102

Young GP, Bush JK (2009) Assessment of the allelopathic potential of Juniperus ashei on germination and growth of Bouteloua curtipendula. J Chem Ecol 35(1):74–80. https://doi.org/10.1007/s10886-008-9585-1

Yuan Z-Y, Li L-H (2007) Soil water status influences plant nitrogen use: a case study. Plant Soil 301(1–2):303–313. https://doi.org/10.1007/s11104-007-9450-y

Zhang GS, Dong Z, Wang LH (1997) Research review of the bioecology characteristics, growth and reproductionof Sbina vulgaris. Neimenggu Forest College Report 19:69–75

Zhang G, Zhang X, Wang L, Liang X, Guosheng W, Yu H (2005a) RAPD analysis on genetic differentiation of Sabina vulgaris populations in Inner Mongolia area. J Arid Land Resour Environ 19(7):193–198

Zhang J, Zhao H, Zhang T, Zhao X, Drake S (2005b) Community succession along a chronosequence of vegetation restoration on sand dunes in Horqin Sandy land. J Arid Environ 62(4):555–566. https://doi.org/10.1016/j.jaridenv.2005.01.016

Zhang G, Wang Z, Wang L, Hao Y, Wen G (2006) Regenerative seedlings dynamics of natural Sabina vulgaris community in Mu Us sandland. Scientia Silvae Sinicae 42(5):62–67 (in Chinese with English summary)

# Chapter 9
# Characteristic Growth Strategies of Trees Planted in Sand Shifting Environments

**Norikazu Yamanaka**

## 9.1 Salicaceae Trees Planted in Shifting Sand Environments

Trees of the Salicaceae family, such as the genus *Salix* and the genus *Populus*, are often used for afforestation to stop sand movement and stabilize sand dunes. Salicaceae trees are highly resistant to the effects of sand movement, especially sand burial, so *Salix psammophila* and *Salix matsudana* are often planted in the Mu Us Sandy Land, and the *Populus simonii* is often planted from the Mu-Us Sandy Land into the adjacent Kubuqi Desert (Fig. 9.1).

These Salicaceae trees are naturally found in areas with good moisture conditions and are not highly drought tolerant. However, due to the high water table in the Mu Us Sandy Land, they can be planted mainly in inter-dune lowlands where moisture conditions are favorable. Below, we will focus on these three species of Salicaceae trees and explain their characteristics and growth conditions in shifting sand dune areas.

First, the most widely planted species in the Mu Us Sandy Land is *S. psammophila*. This species is a shrub-like willow that grows to about 4 m in height and has a bushy tree shape consisting of many thin stems. It naturally grows in sandy lands in Inner Mongolia, Ningxia (Yanchi County), Shaanxi, and Shanxi, and can grow in relatively dry, nutrient-poor sandy lands due to its high environmental adaptability (Flora of China @ efloras.org 2024.11.24 a). For this reason, it is widely used as a greening tree for sand dune stabilization. Especially in the Mu Us Sandy Land, it is planted mainly on the windward side of shifting sand dunes for sand dune stabilization, and it has the effect of lowering the height of shifting sand dunes by suppressing sand movement on the windward side. Also, *S. psammophila* can easily root and establish itself by cuttings. Furthermore, once it takes root, it is known for its strong vitality and shows vigorous growth even in places with large sand movement (Fig. 9.2a). Planting is generally done by direct stem insertion, and in the Mu Us Sandy Land and Kubuqi desert, a method of inserting a group of five cuttings,

 
K. Yoshikawa et al., *Dryland Vegetation in the Sand-Shifting Environment*,
https://doi.org/10.1007/978-981-95-4735-7_9

**Fig. 9.1** Salicaceae trees planted on shifting sand dunes

each about 50 cm long, is often used. It is also used as a material for checkerboards sand barrier for greening sand dune areas. In addition to its use for sand dune stabilization, its leaves and branches serve as livestock feed. Furthermore, its branches are used to weave wicker products and make fences, and, in recent years, it has been used as a raw material for pulp, making it a tree species with a wide range of uses. It has a strong ability to regenerate above-ground parts through sprouting, and even if the above-ground parts are heavily harvested for use, new above-ground parts regenerate the following year (Fig. 9.2b, c).

*S. matsudana* is a tall willow tree that reaches a height of 18 m and a diameter at breast height of 80 cm and is often planted in plains and riverbeds throughout China. Its distribution is wide, and it can be found in Anhui, Fujian, Gansu, Hebei, Heilongjiang, Henan, Jiangsu, Liaoning, Inner Mongolia Autonomous Region, Qinghai, Shaanxi, Sichuan, and Zhejiang provinces in China (Flora of China @ efloras.org 2024.11.24 b). In the Inner Mongolia Autonomous Region of China, it is a tree often used for dune fixation in shifting dune areas. In the Mu Us Sandy Land, the native *S. matsudana* is not seen, but many trees of *S. matsudana* are planted in rows in relatively moist places mainly on interdunal lowlands for dune fixation (Fig. 9.3).

In the Mu Us Sandy Land, where tall trees do not grow, they stand out even from a distance. The planting of *S. matsudana* in the Mu Us Sandy Land is often done by inserting thick branches, about 3 m long and about 10 cm in diameter at the base,

**Fig. 9.2** *Salix psammophila* is often planted on shifting dunes (**a**). Its above-ground parts are harvested in the winter and used for various purposes (**b**). After harvesting, the above-ground parts regenerate through sprouting (**c**)

directly into the sand after removing the leaves and branches. This is a method that takes advantage of the high groundwater level characteristic of the Mu Us Sandy Land and the good rooting ability of *S. matsudana*. The thick branches used for cuttings are cut off from autumn to winter, and the leaves and thin branches are used as winter feed for livestock. The cut branches are soaked in water for about a month before planting to prevent drying and maintain rooting ability. In the spring, the thick branches are inserted into the dunes, usually with branches of 3 m in length, and about 1 m is buried in the sand. After insertion, many adventitious roots grow from the underground part buried in the sand, and many shoots emerge from the aboveground part. In the Mu Us Sandy Land, the branches are repeatedly cut off, mainly on the top part, during the growth process and used for purposes such as livestock feed and fences. Therefore, the many shoots that have emerged grow in clusters, resulting in a tree shape that is completely different from the natural tree shape, and many trees look like upside-down brooms.

**Fig. 9.3** *Salix matsudana* planted on a shifting sand dune. The thick branches are cut off from autumn to winter (**a**). The cut branches are soaked in water before planting (**b**). In the spring, the soaked branches are inserted into the dunes (**c**)

*P. simonii* is a species of poplar that can reach a height of 20 m and a trunk diameter of 50 cm, and its growth range is wide, including mountains, plains, alluvial lands, and valleys. It can also grow in a range of 0–3000 m above sea level and is distributed in Hebei Province, Heilongjiang Province, Jiangsu Province, Jilin Province, Liaoning Province, Inner Mongolia Autonomous Region, Qinghai Province, Shaanxi Province, Shanxi Province, Sichuan Province, Yunnan Province, and Zhejiang Province (Flora of China @ efloras.org 2024.11.24 c). Although its original habitat is mainly on places with relatively good moisture conditions, it is widely planted in the Inner Mongolia Autonomous Region due to its excellent environmental adaptability. Although it is not a common tree species in the Mu Us Sandy Land, it plays a leading role in afforestation of dune lands in the Kubuqi Desert, which is adjacent to the north of the Mu Us Sandy Land. The Kubuqi Desert, like the Muus desert, has a relatively high groundwater level, making it possible to plant trees, and *P. simonii* is planted mainly in places with good moisture conditions

**Fig. 9.4** *Populus simonii* planted on shifting sand dunes. The lower parts of the trunks are buried in sand due to the movement of the sand dunes (**a**). The roots are exposed because the sand has been blown away by the wind (**b**)

**Fig. 9.5** Root sprouting of *Populus simonii*

centered on interdunal lowlands. The growth of planted trees is affected by being buried by sand or having the sand blow away around their roots (Fig. 9.4), but *P. simonii* has high adaptability to such sand movement.

One of the high adaptations to sand movement is the ability of *P. simonii* to generate sprouts from its roots. In addition to roots that extend downward, *P. simonii* has horizontal roots that extend long horizontally, and when the sand on the ground surface is blown away and the horizontally spread roots are exposed, many adventitious buds start to emerge on the surface of the roots. Each bud can eventually grow into an independent individual (Fig. 9.5). This phenomenon is called root sprouting, and due to this ability, *P. simonii* is used as an important dune-fixing tree species.

## 9.2 Dune Environment and the Growth of Salicaceae Trees

### 9.2.1 *Growth Characteristics of* **Salix psammophila** *Growing in Dune Environments*

*S. psammophila* is widely planted, usually on the windward side of the shifting sand dune, in the Mu Us Sandy Lands. We can find young plants and seedlings regenerated on the lower part of the dune slopes and on the interdune lowlands where moisture conditions are favorable. For *S. psammophila* growing on shifting sand dunes, it is very important to cope with sand movement, especially to adapt sand deposition, and we can see *S. psammophila* growing vigorously even though they are buried in sand. Teraminami et al. (2010) conducted a total excavation survey of the aboveground and underground parts of 36 *S. psammophila* seedlings growing on shifting sand dunes to investigate the growing characteristics of *S. psammophila* in an environment of sand deposition. The seedlings of the 36 individuals growing near the foot of a shifting sand dune, with a dune height of about 7 m and a slope angle of about 30°, were selected. The seedlings were divided into three groups, the upper part group (50 cm to 1 m from the lowland surface between the dunes), the middle part group (20–50 cm from the same), and the lower part group (20 cm from the lowland surface), and a growth survey including underground excavation was conducted. The results of the survey showed that all the seedlings examined were 3–4 years old. It was clear that the germination site was not on the surface of the sand dunes where they're currently growing, but on the lowland surface between the hills. Additionally, differences in seedling size and growth were observed depending on their position on the dune slope. Seedlings growing at higher positions on the slope were larger than those growing at lower positions (Fig. 9.6).

Since the age of the seedlings studied and the position of germination are almost the same, the difference in seedling size due to the different position on the dune slope reflects the difference in depth of burial in the sand. Specifically, when examining the relationship between the distance from the germination point to the surface of the dune slope and the length of the above-ground part, it was found that the longer the distance to reach the dune surface, the longer the above-ground part (Fig. 9.7). Also, regarding the morphology of the underground part, it is considered that *S. psammophila* has the ability to increase the amount of the underground part by being buried in sand, thereby increasing the amount of aboveground growth, as the main stem length of the buried part is larger for seedlings growing on the upper part of the sand dune (Teraminami et al. 2010).

Furthermore, as a comparative study with *S. psammophila*, a survey was conducted on the impact of sand burial on the growth of *Salix cheilophila*, a shrub willow that grows naturally in the same Mu Us Sandy Land. In the results, unlike *S. psammophila*, no correlation was found between the length of the buried part of the main stem and the biomass and the number of adventitious roots in the case of *S. cheilophila*, suggesting that *S. cheilophila* has poor adaptability to sand burial (Teraminami et al. 2013).

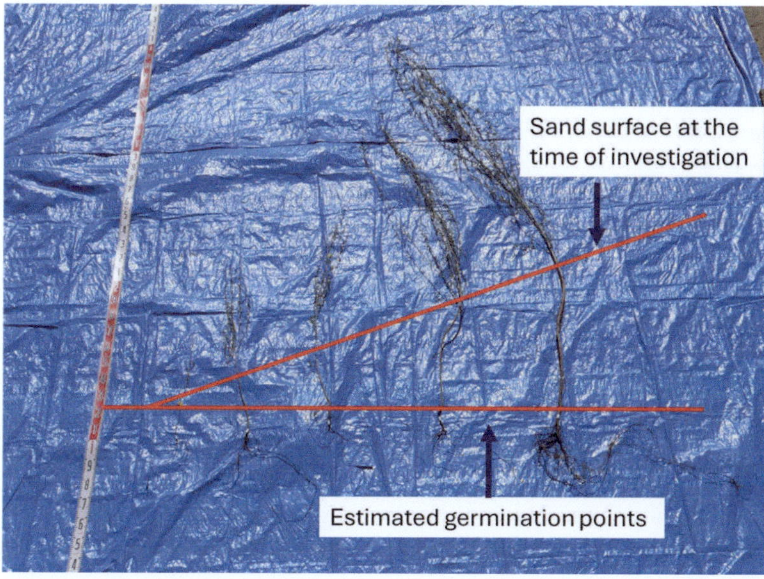


**Fig. 9.6** Estimated germination points of *Salix psammophila* seedlings (4 years old) dug from the different elevations on the dune slope. (Photo by A. Nakashima, Wakayama Univ.)

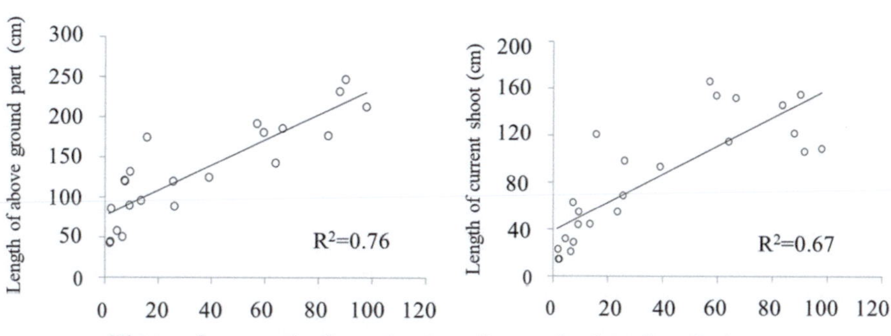


**Fig. 9.7** Growth of *Salix psammophila* seedlings (4 years old) dug from the different elevations on the dune slope. (Modified from Teraminami et al. 2010)

Thus, even if they are shrub willows growing in the same Mu Us Sandy Land, it is very interesting that the adaptability to sand movement, especially sand burial, is different, and in the Mu Us Sandy Land, *S. psammophila* 's high resistance to sand burial is well utilized, and planting is being carried out in shifting sand dune areas where sand movement is intense.

### 9.2.2 *Adaptation of* Populus simonii *Planted in the Kubuqi Desert to Sand Movement*

*S. psammophila* and *S. matsudana* have strong resistance to sand burial. Similarly, *P. simonii* also has strong resistance to burial by sand. A study of the growth of *P. simonii* growing on the slopes of sand dunes in the Kubuqi Desert, about 7 years after planting, showed that trees growing on the slopes of sand dunes while remaining buried in sand grew more vigorously compared to trees growing on flat ground without being buried in sand (Fig. 9.8). The trees growing in the buried sand environment were also larger (Miyazaki et al. 2014a).

When the root system of *P. simonii* growing on the slope is examined, like *S. psammophila* and *S. matsudana*, vigorous adventitious roots are produced from the buried trunk, which is thought to support the vigorous growth in the sand-depositing environment (Miyazaki et al. 2014a; Yamamoto et al. 2015). Furthermore, it is often observed that each branch grows independently by producing adventitious roots from the buried branches, eventually showing growth in the form of a clonal population with a cluster structure (Yamanaka and Yamamoto 2016), which is also considered to support the vigorous growth in the sand-depositing environment.

Thus, *P. simonii* shows vigorous growth in a sand-depositing environment, but it can also grow in an environment where the sand around the root system is blown away by the wind and the roots are exposed. In normal plants, if the sand around the roots is blown away and the roots are completely exposed, it leads to death. In the case of *P. simonii*, if the roots are rapidly exposed immediately after planting, growth becomes difficult and often leads to death, but if the roots are exposed when the horizontal roots that spread widely in the horizontal direction are sufficiently developed, many adventitious buds are formed on the horizontal roots, and these

**Fig. 9.8** Growth status of *Populus simonii* about 7 years after planting

buds develop and grow as new individuals as root sprouts. This formation of adventitious buds on horizontal roots does not occur deep underground but is promoted in environments where sand is blown away and light hits the horizontal roots. *P. simonii* is a tree species that can produce new individuals through growth and root sprouting not only in environments where sand accumulates but also in environments where sand is removed and roots are exposed, and it can be said to have a strategy to flexibly respond to environmental changes associated with sand movement.

## 9.3 Comparison of Sand Burial Resistance of Salicaceae Trees Through Sand Burial Experiments

The high sand burial adaptability of *S. psammophila*, a shrub willow widely used for sand dune fixation in the Mu Us Sandy Land, has been clarified in field surveys, but here we explain the sand burial resistance of *S. matsudana*, a tall willow used for sand dune fixation in the same Mu Us Sandy Land. The following is the result of a comparative planting experiment with *P. simonii* and *Populus alba*, which are tall poplar species used for dune fixation in the same Mu Us Sandy Land and in the adjacent Kubuqi Desert.

Miyazaki et al. (2014b) analyzed the effect of artificial sand burial on the growth of cuttings through a planting experiment with tall Salicaceae plants. The experimental materials were 1-year-old cuttings of *S. matsudana*, *P. simonii*, and *P. alba*. Cuttings were grown in an environment where cuttings were buried in sand at 0, 50, and 100% of their height at the beginning of the experiment and examined the effect of the degree of burial on growth. As a result of the experiment, it was confirmed that the growth of all three species rapidly increased stem length in the 100% burial group. *P. alba* showed a significant increase in elongation growth even at 50% burial, but no difference in elongation growth was observed between the 0% and 5% groups for *P. simonii* and *S. matsudana* (Fig. 9.9).

Significant promotion of diameter growth was also observed in the 100% burial group for all three species. However, no difference was observed between the 0% and 5% groups in diameter growth for all three species. Furthermore, in terms of total plant biomass production, *P. alba* did not show a significant increase in biomass due to burial, but *S. matsudana* and *P. simonii* showed a significant increase in biomass growth due to burial, especially 100% burial, revealing their excellent adaptability to burial. This growth promotion by burial is mainly brought about by the formation of adventitious roots in the underground part and the accompanying increase in the total amount of roots, and it is thought that the role of ethylene, a plant hormone involved in the formation of adventitious roots, has a large influence (Miyazaki et al. 2014a). Similar results to Miyazaki's planting experiment using cuttings were also observed in *P. simonii* planted in the Kubuqi Desert, where trees buried in sand grew larger (Tateishi et al. 2013). Additionally, *S. matsudana* trees in

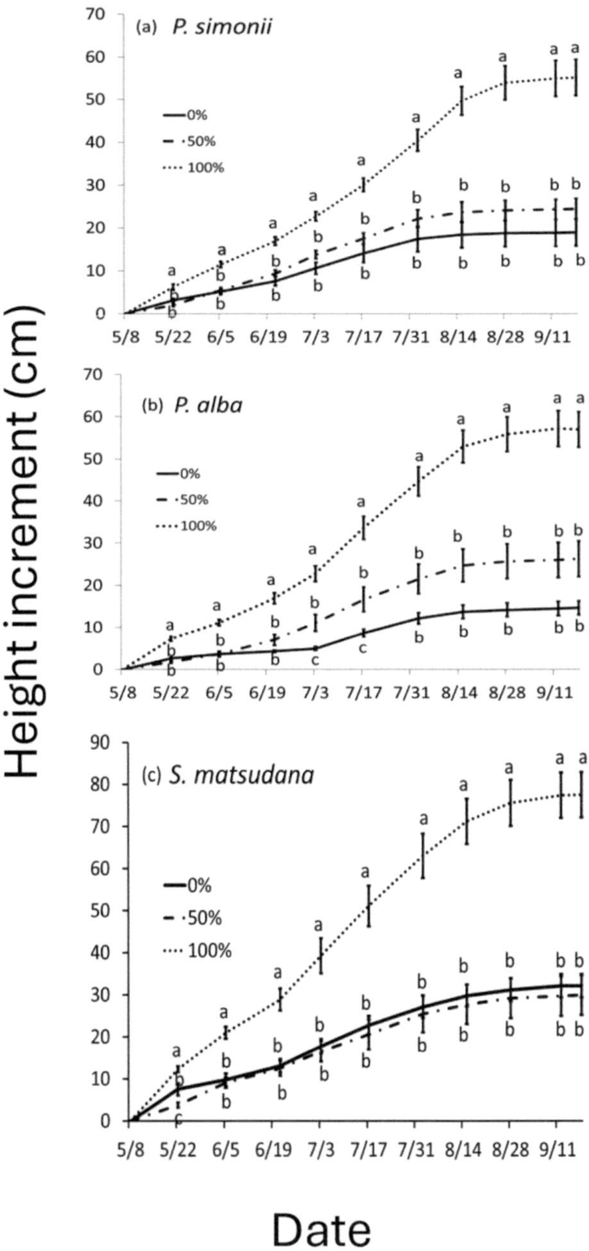


**Fig. 9.9** Changes in stem height growth of cuttings at 0%, 50%, and 100% burial depth, (**a**) *Populus simonii,* (**b**) *Populus alba,* (**c**) *Salix matsudana*. Error bars indicate standard error (SE). (Reproduce from Miyazaki et al. 2014b)

the Mu Us Sandy Land are known to develop numerous adventitious roots from the buried trunk when buried in sand (Masuda et al. 1988: see Fig. 9.10). The rapid development of adventitious roots after burial and the subsequent larger root system are thougtht to promote the growth of buried trees. This rapid development of adventitious roots is an important characteristic for tree species used in afforestation within shifting sand environments.

**Fig. 9.10** Adventitious roots emerging from the trunk of a *Salix matsudana* tree buried in sand. (Photo by K. Yoshikawa, Okayama Univ.)

## References


Flora of China @ efloras.org (Viewed on 2024.11.24 a) *Populus simonii* (www.efloras.org/flora-taxon.aspx?flora_id=2&taxon_id=200005709)

Flora of China @ efloras.org (Viewed on 2024.11.24 b) *Salix matsudana* (www.efloras.org/flora-taxon.aspx?flora_id=2&taxon_id=200005909)

Flora of China @ efloras.org (Viewed on 2024.11.24 c) *Salix psammophila* (www.efloras.org/florataxon.aspx?flora_id=2&taxon_id=200005972)

Masuda T, Kobayashi T, Yoshikawa K, Morimoto Y, Kobashi S (1988) The root distribution of *Salix matsudana* in the Mu Us Desert, China. Bull Reveget Res 10:42–55 (in Japanese)

Miyazaki K, Okada Y, Tateishi M, Yamamoto F, Mao H, Taniguchi T, Yamanaka N (2014a) Effects of sand burial on growth and osmotic adjustment in Simon poplar (*Populus simonii* Carr.) planted in Kubuqi desert, Inner Mongolia, China. J Jpn Soc Reveget Technol 40:31–36 (in Japanese with English summary)

Miyazaki K., Okada Y., Yamamoto F., Mao H., Taniguchi T., Yamanaka N. (2014b) Effects of artificial sand burial on growth of *Populus simonii*, *P. alba* and *Salix matsudana* cuttings. J Jpn Soc Reveget Technol 40:37–42 (in Japanese with English summary)

Tateishi M, Miyazaki K, Yamamoto F, Mao H, Okada Y, Yamanaka N (2013) Effects of sand burial on water use and growth of Simon poplar (*Populus simonii* Carr) in Kubuqi desert, Inner Mongolia of China. J Jpn Soc Reveget Technol 39:68–73 (in Japanese with English summary)

Teraminami T, Nakashima A, Nakazato N, Ichihara S, Kurotaki M, Yamamoto M, Yamamoto M, Yoshikawa K, Zhang GS (2010) Effects of pile up of sand on growth and morphology of *Salix psammophila* in the Muus sandy land, Inner Mongolia, China. J Jpn Soc Reveget Technol 36:87–92 (In Japanese with English Summary)

Teraminami T, Nakashima A, Ominami M, Yamamoto M, Zhang GS, Yoshikawa K (2013) Effects of sand burial depth on the root system of *Salix cheilophila* seedlings in Mu Us Sandy Land, Inner Mongolia, China. Landsc Ecol Eng 9:249–257

Yamamoto F, Fujiwara K, Taniguchi T, Mao H, Yamanaka N(2015) Effects of sand burial or erosion on the growth and physiology of *Populus simonii* trees in Kubuqi desert, China. J Jpn Soc Reveget Technol 41:39–44 (in Japanese with English summary)

Yamanaka N, Yamamoto F (2016) Ecosystem restoration in China. In: Kurosaki Y, Kurozawa Y, Shinoda M, Yamanaka N (eds) Asian dust – effects on human health, environment, and counter measures. Maruzen Publishing Co. Ltd., Tokyo, pp 133–138 (in Japanese)

# Part V
# Desertification/Land Degradation as a Global Environmental Issue

## Introduction

Land degradation is the deterioration of the capacity of land to provide ecosystem productivity, functions, and services. Such land degradation occurs not only in drylands but also in humid areas, and due to prolonged climate change and inappropriate land use in various places, severe land degradation is expanding worldwide.

Especially in drylands, the combination of climate change and inappropriate human activities leads to many adverse consequences, such as degradation and loss of ecosystems and sandstorms that threaten human life. In recent years, there has been an increase in cases where sandstorms originating in desert areas have swept into modern cities in northwestern China, Africa, the western United States, Australia, etc. As desertified lands expand, they affect human survival and socioeconomic development by reducing habitable areas and destroying production bases (Qi et al. 2015).

However, it is not easy to identify drylands, the areas where desertification is likely to occur. Furthermore, due to the regional specificity of desertification progression and the spatiotemporal diversity of land degradation phenomena caused by differences in internal environmental conditions, the actual situation of land degradation in drylands has not always been well understood. Nevertheless, since land degradation in drylands is a major environmental issue facing terrestrial ecosystems on Earth, it is necessary to quantitatively evaluate the progression of desertification and implement effective countermeasures. However, the urgency of combating desertification depends on values of dryland ecosystems, which makes the assessment of its progress and recovery processes subjective, and objective and common assessment methods have not yet been standardized (Fig. 1).

**Fig. 1** Sandstorms frequently occur in dryland. A herdsman sets out on horseback to retrieve a stray sheep from the pasture

This chapter will explain the background of desertification/land degradation based on scientific research results, and consider the current status and future prospects of measures against desertification/land degradation.

## Reference

Qi F et al (2015) What has caused desertification in China? Sci Rep 5. https://doi.org/10.1038/srep15998

# Chapter 10
# Current Status and Background of Desertification/Land Degradation in China

**Ken Yoshikawa and Linhe Wang**

Desertification/land degradation progressing in China is an environmental issue that requires strong measures because it mercilessly destroys the foundation of people's livelihoods. The main causes include climate change and human activities, such as land-use changes from agricultural development and urbanization, as well as inappropriate land management practices like unplanned fuelwood harvesting. In other words, we are destroying our own livelihood base and endangering our own survival through our own activities.

The degradation and extinction of *Juniperus sabina* (=*Sabina vulgaris*) communities in the Mu Us Sandy Land is land degradation in shrub forest lands, which is the loss of function of the forest ecosystem. The degradation of forest ecosystems in drylands causes further degradation of the environment over a much wider range than in humid lands. Reafforestation in drylands as a measure against desertification is urgently required, but there are not many tree species that can be used as restoration materials. By clarifying the ecophysiological characteristics of as many tree species as possible to understand the species specificity of the tree species growing in drylands, it is indispensable to expand the range of choices for reafforestation materials for the proper restoration of degraded drylands.

## 10.1 Causes of Desertification/Land Degradation

### *10.1.1 Wind Erosion*

China is one of the countries most severely affected by desertification/land degradation. The agricultural-pastoral boundary zone in northern China is susceptible to wind erosion due to its arid conditions, the surface being covered with porous sandy deposits, minimal vegetation, and concentrated winds during the dry season from winter to spring (Fig. 10.1). In China, 70.2% of the desertified/degraded land is due

 
K. Yoshikawa et al., *Dryland Vegetation in the Sand-Shifting Environment*,
https://doi.org/10.1007/978-981-95-4735-7_10

**Fig. 10.1** Wind erosion. In dryland, sparse vegetation allows strong winds to sweep across the surface

to wind erosion, while less than 16% of land degradation is due to water erosion and salt accumulation (Zan et al. 2023). In other words, desertification in China is "sandy" desertification, and the recovery of vegetation on sandy soil is the most crucial measure against desertification.

Whether sand dunes move or not is determined by the balance between the wind erodibility of sand surface related to vegetation coverage and the sand transport capacity of the wind. For example, when farmland is abandoned and vegetation cover decreases to 20–30%, the underlying sand dunes reactivate and gradually merge to form larger desert patches (Warren and Agnew 1988). The wind near the ground surface is directly related to soil erosion, and high wind activities are strongly associated with high risk of desertification.

Forests can reduce wind velocity and are often referred to as green barriers. In the Mu Us Sandy Land in 2000, only local areas in the southern and northeastern areas had a vegetation cover rate (VCR) over 40%, while more than half of the area had a VCR less than 10%, but by 2020, only about 10% of the area had a VCR less than 10%, and more than half of the area had a VCR over 40% (Wang et al. 2022). This is thought to be the result of the creation of windbreak forests through afforestation, which reduced wind speed and promoted the growth of surrounding vegetation. Particularly, when the wind speed from April to July is lower than 2.46 m/s, the VCR in the Mu Us Sandy Land is expected to increase by 5%. The reduction of extremely desertified land from over 60% in 2000 to about 15% in 2020 is likely achieved through the decrease in wind speed.

### 10.1.2 Global Warming

Meanwhile, strong warming is occurring throughout China, and compared to 1961, the average temperature everywhere has increased by more than 2 °C, which is equivalent to 150 mm in the potential evapotranspiration. Particularly, the average temperature per decade has risen by 0.41 °C in the northern and northeastern regions and 0.38 °C increased in the northwestern region (Lyu et al. 2020).

Changes in precipitation due to global warming have been observed since the 1980s, and from 1961 to 2019, the annual average precipitation in arid and semi-arid regions generally showed a weak increasing trend. However, there were large differences in the amount of precipitation between regions, and the trend was not sufficient to explain the changes in desertification area during that period. This is because global warming does not change precipitation in the same way everywhere; some areas become warm and humid, while others become warm and dry.

For example, in the northern and northeastern regions, precipitation has been decreasing by 10.0 mm every 10 years, leading to increased dryness, and serious droughts have occurred several times in recent decades. The drought in the late 1990s led to frequent sandstorms, and in the northern region, natural disasters such as wind erosion, sand drift, desertification, and dust storms continue due to drought. On the other hand, in the northwestern region, precipitation is higher than in the northeastern region, and particularly, precipitation in winter and spring has increased by 6.8 mm every 10 years since 1961. Particularly, spring rainfall have promoted plant growth, which in turn preventing the severity of wind erosion, further affecting the vegetation restoration on the periphery of the desert and the recovery of desertified lands.

Changes in precipitation can increase or decrease vegetation, but even if rainfall increases and drought is alleviated, land degradation may progress due to increased soil runoff caused by rainfall (Fig. 10.2). For example, in the upper and middle reaches of the Yellow River, where precipitation is increasing, the amount of sediment flowing into the Yellow River from the Ulan Buh Desert is increasing, with the average of sediment transport at Tuodaoguai (where the middle reaches begin) increasing by 28.3% in 10 years.

### 10.1.3 Human Activity or Natural Factors?

The process of desertification is complex, with both human and natural factors (such as climate) acting together, and often interacting in complicate manner (Qi et al. 2015). Therefore, it is virtually impossible to separate the effects of the two. The complexity of the effects of anthropogenic and natural factors on desertification can be observed throughout northern China. For example, from the 1950s to the recent recovery period, the pressure on land degradation has continuously increased due to population growth and economic development (Fig. 10.3), so assuming that

**Fig. 10.2** The Mu Us Sandy Land after heavy rain. Rainwater has pooled in the interdune depressions, particularly in areas closer to the leeward side of the shifting sand dunes

**Fig. 10.3** Downtown of Datong City (Shanxi Province), developed in dryland. With its growing population, the town is bustling with activity

those anthropogenic factors are the main cause of desertification does not explain the recovery in land degradation observed since 2000. As a result, some studies have concluded that climate change has affected the soil quality, vegetation cover, species composition, and hydrologic cycles in drylands, leading to the expansion of desertified areas, while other studies have suggested that unsustainable traditional land use such as pasturing, logging, and exploitation of underground water are accelerating desertification. For example, since 1949, the number of livestock in the northwestern China has shown an increasing trend, exceeding the rational carrying capacity of this region by 33.6% in 2008, so it is suggested that the degradation of grasslands due to soil erosion and sand accumulation is due to stocking pressure caused by human activities.

The land degradation factors affect dryland environment in different ways depending on spatial and temporal scales. For example, climate change is the main driving force of land degradation through the geological ages while human interference is local (small area) and can cause grassland degradation or restoration and reactivation of sand dunes in a short period of time. Even lands that have been under severe stress from aridification for decades can experience human-induced desertification if excessive human activities or management mistakes lead to widespread land degradation in a short period of time. In other words, when the spatiotemporal scale is reduced, it is easier to conclude that human activities are strongly involved in the desertification process.

## 10.2 Background of Desertification/Land Degradation in the Mu Us Sandy Land

Based on such a background, most previous studies have recognized the anthropogenic factors (population increase, unsuitable reclamation, and overgrazing) as the main driving forces of the rapid desertification in the Mu Us Sandy Land before the 1990s, while natural impacts, such as global warming, as a minor driver (Zhou et al. 2015). Therefore, the Mu Us Sandy Land is well known worldwide, not just in China, as a "human-made" desert (Wang et al. 2008). From an extreme perspective, the serious desertification that occurred in this area has been attributed primarily to human activities over the past 2000 years. However, according to ancient Chinese historical records and archaeological evidence, the modern Mu Us Sandy Land existed at least 4000 years ago, long before large numbers of people lived in this region, and from that time to the present, it has experienced several phases of desertification and rehabilitation that paralleled climatic changes.

On the other hand, most of the Mu Us Sandy Land has precipitation over 300 mm per year, which can support a vegetation cover as is permitted by the natural conditions. There should not be so large in area of shifting sand dunes, which takes up approximately 45% of the region, unless it is caused by some human-induced disturbances. Therefore, the rapid desertification since at least the 1950s is considered the result of irrational cultivation and overgrazing.

**Fig. 10.4** Sheep and goats in the Mu Us Sandy Land

The rate of expansion of the southern edge of the Mu Us Sandy Land was not as much as the several km per year that the Chinese media exaggeratedly reported as the "threatening invasion of the Mu Us Sandy Land" from the 1980s to the 1990s, but according to satellite images from that period (1978–2007), small sand dunes about 100–200 m in size were swallowing up grasslands at the southeastern front at a rate of about 10–20 m per year (Wu et al. 2008).

Meanwhile, the number of livestock (sheep and goats) and the human population have been increasing over the past 50 years (Fig. 10.4). In addition, from the mid-1950s to the mid-1990s, over-reclamation occurred in this region, but few areas of arable land became desertified during these periods, and significant rehabilitation occurred after the 1980s despite these increases. Therefore, the impact of human activities has both destructive effects, such as deforestation and land abandonment, and restorative effects, such as cultivation and greening. The southeastern part of the Mu Us Sandy Land, where farmland and pasture are mixed together, is slower in the expansion of desertification than the central-western parts where there is only pasture, because the cultivated farmland is protected from desertification by human intervention. Therefore, while human activities may have had some impact on the desertification of this region, they also contribute to the recovery and conservation of land degradation, suggesting that overall, they do not strongly contribute to desertification, and climate change is considered an important factor in the desertification of the Mu Us Sandy Land.

## References

Lyu Y et al (2020) Desertification control practices in China. Sustainability 12. https://doi.org/10.3390/su12083258

Qi F et al (2015) What has caused desertification in China? Sci Rep 5. https://doi.org/10.1038/srep15998

Wang X et al (2008) Desertification in China: an assessment. Earth Sci Rev 88:188–206

Wang X et al (2022) Desertification in the Mu Us Sandy Land in China: response to climate change and human activity from 2000 to 2020. Geogr Sustainability 3:177–189

Warren A, Agnew C (1988) An assessment of desertification and land degradation in arid and semi-arid areas, Drylands Programme Paper 2. International Institute for Environment and Development, University College

Wu W et al (2008) Land degradation monitoring in the west Muus, China. Int Arch Photogramm Remote Sens Spat Inf Sci 37:847–858

Zan GS et al (2023) Key data results and trend analysis of the sixth national survey on desertification and sandification. Forest Resources Management. https://doi.org/10.13466/j.cnki.lyzygl.2023.01.00

Zhou et al (2015) Long-term vegetation changes in the four mega-sandy lands in Inner Mongolia, China. Landsc Ecol 30:1613–1626

# Chapter 11
# Global Impact of Desertification/Land Degradation

**Ken Yoshikawa**

## 11.1 Impact of Desertification/Land Degradation

The globalization of the world economy has created environmental issues that were hardly conceivable 20 years ago. For example, the increased demand for agricultural products in developed countries (such as Japan) is causing deforestation for farmland expansion in developing countries (such as Brazil). This impact of such local activities in remote areas is called telecoupling. Furthermore, telecoupling is not limited to the movement of products between two regions. For example, when corn is exported, the water required to produce it is also considered exported. In telecoupling caused by increased demand in the consumption area, it is difficult to notice the causal relationship between land degradation in remote areas and the imported consumer goods at hand, and people in developed countries hardly think that their activities are causing land degradation in remote areas (Cherlet 2018) (Fig. 11.1).

Desertification/land degradation affects the ecosystems of a wide surrounding area, so if the target area is limited to drylands, the impact of land degradation cannot be accurately assessed. Therefore, the issue of desertification/land degradation is a global problem that cannot be confined only to drylands, and it should be addressed in all regions of the world (Cherlet 2018).

By defining the applicable sites as "areas affected by land degradation in drylands" as defined in the United Nations Convention to Combat Desertification (UNCCD) articles, it becomes consistent with Land Degradation Neutrality (LDN). LDN refers to a state where the quantity and quality of land resources necessary to support ecosystem functions and services and enhance food security are stable or increasing within a specific temporal and spatial scale, but not restricting the spatial extent. In other words, without limiting the site conditions such as drylands, LDN aims to maintain or enhance the land-based natural resources and the ecosystem services derived from it, through the three perspectives: biological productivity, ecological integrity, and value to humans in all relevant areas (IPCC 2019).



K. Yoshikawa et al., *Dryland Vegetation in the Sand-Shifting Environment*,
https://doi.org/10.1007/978-981-95-4735-7_11

**Fig. 11.1** Rainfed farmland for maize cultivation in Kenya

While the establishment of LDN is being promoted, it is estimated that the world's crop yield will decrease by about 10% by 2050 due to land degradation and climate change. The regions where production is decreasing are India, China, Sub-Saharan Africa, etc., which export many crops to developed countries, and the decline in overall production makes it difficult to realize the LDN. All developed countries that import a lot of food and materials from these regions can no longer ignore this problem.

## 11.2 Area of Desertified Land

Since Stebbing's pioneer study in 1935, a variety of estimates have been published regarding the area of desertified land. Some studies have provided astonishing figures both the rate of advance of desert lines and the area affected by desertification (Lamprey 1975; UNEP 1984). To organize such situations and ensure the implementation of UNCCD, the first edition (UNEP 1992) and the second edition (UNEP

1997) of the World Atlas of Desertification (WAD) and the Millennium Ecosystem Assessment report (MA), published in 2005, have received a certain level of recognition worldwide and are important documents for understanding the current situation of desertification. Furthermore, about 20 years after the second edition, considering the development of remote sensing technology for land information and the increase in human activities worldwide, the third edition of the WAD was published in 2018 (Cherlet 2018). However, unlike previous editions, it does not include maps indicating the distribution of desertified land in the world. In SRCCL, published the following year (2019), although a great deal of data on land resources is currently available, the estimation of the current potential production in degraded land is greatly hindered by this unreliable information due to the lack of understanding of the location, area, and condition of the degraded land. In fact, there is no clear consensus on the extent of degraded land, not only worldwide but also within specific countries. While land degradation is a serious issue, it has not been possible to get a clear picture of the states of land degradation at any level, such as regional or national levels (Fig. 11.2).

As previously mentioned, estimation of the area and degree of change in land degradation varies significantly depending on the estimation method (Gibbs and Salmon 2015). For example, Global Assessment of Soil Degradation (GLASOD), commissioned by UNEP, estimated the area of degraded land based on various data and found nearly 20 million $km^2$ of farmland, pasture, forest, and woodland have degraded since the mid-twentieth century. Since GLASOD relied on local knowledge rather than measurement, Bot et al. (2000) developed country-specific data

**Fig. 11.2** Slash-and-burn farmlands prior to crop planting (Madagascar). Ash covers the surface but cannot be distinguished from degraded fire scares

from GLASOD to determine the land area affected by degradation and found that 66% of the world's land, or 61.4 million km$^2$, was degraded. A report using a climate data set from 1981 to 2010 reported that the total degraded land was 63.4 million km$^2$ (Koutroulis 2019). On the other hand, when estimated from FAO data using a biophysical model, the area of land degradation was nearly 10 million km$^2$. In other words, as a result of estimating by various methods such as expert's regional information, satellite observation, biophysical models, and taking inventory of abandoned farmlands, the estimated area of degraded land ranged from less than 10 million km$^2$ to over 60 million km$^2$. Moreover, their spatial distribution was also significantly different.

The reasons for this difference in estimates include the lack of a consistent standard for the process of desertification/land degradation (what kind of degradation phenomena are due to desertification), the time scale of desertification (how long ago did desertification start), and the method for recognizing desertified land (what is the indicator for assessing degradation). This lack of standards can lead to arbitrary judgments.

The recognition of desertification began in 1992 when the UNCCD was adopted at the United Nations Conference of Environment and Development (UNCED) held in Rio de Janeiro, so called "Earth Summit." At the summit, some attempts were made to redefine desertification, advance the recognition of the current state of desertification, and implement countermeasures. However, in the process, it became apparent that there were challenges in identifying the areas where desertification is occurring and in evaluating the situation.

The term "degradation" is an umbrella that includes various states of land (Gibbs and Salmon 2015). For example, desertification, salinization (Fig. 11.3), erosion (Fig. 11.4), compaction, and invasion of alien species, the phenomena are diverse. Moreover, according to the definition of desertification, land degradation is attributed to climate change and human activity, but both have diverse processes of degradation, and it is impossible to monitor all processes and phenomena, and it is difficult to identify the causes of individual degradation phenomena.

Land degradation is considered to be a decrease in the productivity of land and soil. However, when focusing on the temporal and spatial scale of this process, the environmental conditions and the length of the process leading to degradation have caused a lot of confusion in the evaluation of the degree of land degradation. In fact, some estimates of degradation focus on the final state of the land, while others take into account the ongoing process of degradation itself, and even the perceived risk of degradation, adding more confusion to the term. For example, both deforested forests and eroded grasslands are considered the same degraded areas, adding further confusion to this term. Another issue is that lands with inherently low productivity, such as wetlands and natural salt deposits, may also be included in degraded land. Furthermore, while most seminal estimates have focused on soil degradation, more recent efforts have been investigating broader issue of land degradation from an ecosystem approach that encompasses both soil and vegetation. As a result, the evaluation has become even more complex.

**Fig. 11.3** Salinization (UAE). In dryland, when groundwater or irrigation water with high salt concentrations evaporates from the surface, salts are left behind, forming a white layer of salt crust over farmland

**Fig. 11.4** Water erosion (Naiman, Inner Mongolia Autonomous Region). In drylands with sparse vegetation, rainwater flowing down gentle slopes collects to form an erosional valley, which gradually expands

The population will continue to increase, and the world population is expected to exceed nine billion by 2050. As a result, there will be a shortage of farmland to meet the increasing demand for food, fiber, and energy. Currently, the world's crop cultivation areas, where crops and fruit trees are harvested, have been created by converting forests (63%) and grasslands (37%) and have increased from 2.65 million km$^2$ in 1700 to 7.7 times that in 2016, covering 13.7% of the world's land. This is most of the land best suited for agriculture, and even if it can be expanded in the further, it is only up to 16.3% of the world's land, leaving little land left. Under such circumstances, degraded lands have often been considered as promising sites to meet the growing global demand for agricultural products, solving issues of land scarcity. However, newly reclaimed dryland is of low quality as a crop cultivation area and is prone to degradation, so it cannot produce as expected. Moreover, if the area of desertification/land degradation is overestimated, there is a risk that policies will be implemented based on future projected values of agricultural productivity that depend on the productive capacity of degraded dryland that does not exist.

## 11.3 Variety of Degradation Factors by Region

Information on desertification/land degradation is limited to certain regions, and there are no consistent criteria for determining degradation. Therefore, it is difficult to understand the state of desertification/land degradation over a wide area, and the WAD (3rd) (2018) does not provide a map of desertification/land degradation for the entire planet, nor does it describe the area of desertification/land degradation. Environmental restoration and conservation measures in each region require the extraction of region-specific factors and the selection of appropriate countermeasure technologies. Therefore, integrating individual information based on a vast amount of environmental data for a specific target area formulating current understanding and countermeasures adopted in the WAD (3rd) can be evaluated as one significant attempt. While the aspects of land degradation vary depending on the circumstances of each region, to understand the trends of environmental degradation and loss of land productivity in the world as a whole or in a particular region, the global aggregation of information as shown in the first and second editions of WAD is effective. Therefore, in order to grasp the concept of desertification/land degradation and to recognize the danger of its progress, there are significant benefits in showing the issue of desertification/land degradation on a global scale with specific figures and location maps, even if they contain large errors. Thus, let us carefully look at the reported area and distribution information of desertification and land degradation (Fig. 11.5).

According to the WAD (1st) (1992), approximately 18% of drylands, including hyper-arid regions, are experiencing land degradation. Among those, about 45% are moderately degraded, while only about 12% is undergoing more severe degradation. When comparing four regions within drylands, the degree of degradation is

**Fig. 11.5** Land reclamation for irrigated farmland in Kenya. Water is drawn via irrigation channels from an exotic river flowing through dryland, transforming grassland into irrigated farmland

greater in more humid regions than in dry regions. This indicates that rainfed and irrigated agricultural lands in relatively wetter semi-arid or dry sub-humid regions are more prone to land degradation than livestock farming widely practiced in dry pastoral grasslands. The land use patterns in drylands differ according to the degree of aridity, and the diverse geographical environments along with various human activities adapted to them are strongly reflected in the progression of land degradation, indicating that the progress of desertification/land degradation is not uniform even within drylands.

In WAD (2nd) (1997), UNEP showed the area of desertification/land degradation by continent and stated that 20% of drylands worldwide were degraded. In particular, degradation was progressing more in Africa, Asia, and Europe. Although the reliability of these figures is low, reflecting the current situation of each continent, different factors, such as severe environmental conditions in Africa and intensive land use in Asia (Figs. 11.6 and 11.7) and Europe, are responsible for land degradation. It has been suggested that the diversity of physical and social factors within drylands is significantly related to the progression of desertification/land degradation.

Every time statistical data is updated, the area of desertification/land degradation becomes smaller and seems to be getting closer to the actual situation. However, in the WAD (3rd), neither maps nor the size of degraded land is shown, as it is difficult to calculate the area of desertification/land degradation uniformly in the same way everywhere in the world. This indicates that it is not easy to identify individual

**Fig. 11.6** Parts of the interdune depression in the Mu Us Sandy Land have been submerged by heavy rainwater, compelling livestock to feed in the water

**Fig. 11.7** Grassland degraded by vehicle ruts. Driving over ruts causes vehicles to sway, creating numerous new ruts across the grassland

desertification phenomena. Nevertheless, the WAD (3rd) reports that the productive capacity of land, assessed in terms of net primary production above the ground, has declined by about 20.4% globally between 1999 and 2013, showing recognition of the declining trend in land productivity.

## References

Bot AJ, Nachtergaele FO, Young A (2000) Land resource potential and constraints at regional and country levels. FAO, p 118

Cherlet M (ed) (2018) World Atlas of desertification, 3rd edn. Publication Office of the European Union, p 237

Gibbs HK, Salmon JM (2015) Mapping the world's degraded lands. Appl Geogr 57:12–21

IPCC (2019) Climate Change and Land. Special report on climate change, desertification, land degradation, sustainable land management, food security, and greenhouse gas fluxes in terrestrial ecosystems. p 896

Koutroulis AG (2019) Dryland changes under different levels of global warming. Sci Total Environ 655(10):481–511

Lamprey HF (1975) Report on the desert encroachment reconnaissance in Northern Sudan, 21 October to 10 November 1975, vol 17. UNESCO/UNEP, Paris/Nairobi; republished in Desertification Control Bulletin, pp 1–7

Stebbing EP (1935) The encroaching Sahara. Geogr J 85:506–524

UNEP (1984) The state of the environment 1984. UNEP

UNEP (1992) World atlas of desertification, 1st edn. Arnold, p 69

UNEP (1997) Atlas of desertification, 2nd edn. Arnold, p 182

# Chapter 12
# Prospects for Combating Desertification/Land Degradation

**Ken Yoshikawa**

## 12.1 Measures to Combat Desertification/Land Degradation

The fragility of the dryland ecosystems in northern China has historically been exacerbated by large-scale reclamation and land misuse policies, with little consideration given to government land management, leading to desertification (Chen and Tang 2005). As a countermeasure, following UNCED (1977), the Chinese government launched five more proactive approaches to prevent and control desertification (Lyu et al. 2020). These are the Three North Shelterbelt Development Program (TNSDP, from 1978 to 2050) (Fig. 12.1), the Conversion of Cropland to Forest Program (CCFP, since 1999), the Natural Forest Protection Program (NFPP, since 2000), the Beijing-Tianjin Sand Source Control Program (BSSC, since 2001) (Fig. 12.2) and the Return Grazing land to Grassland Program (RGGP, since 2003). All programs are forcing to abandon the cultivation of once-reclaimed farmland or enclose pastures to prohibit grazing and aim to restore degraded lands through afforestation and pasture sowing.

Since 1990, despite a significant increase in population density in the Mu Us Sand Land, the policy of "grazing prohibition, grazing rest, and rotational grazing" has been restoring pasture vegetation, which is an effective way to prevent desertification, but it simultaneously deprives the local residents of their sole means of livelihood. Environmental issues cannot be separated from economic activities, and attempting to address environmental issues without considering demographic and social issues is unlikely to succeed. Moreover, since there is no prospect of a reduction in the population pressure on land use in the near future, how to reconcile the livelihoods of local residents with the sustainable land use will be the key to desertification control.

 
K. Yoshikawa et al., *Dryland Vegetation in the Sand-Shifting Environment*,
https://doi.org/10.1007/978-981-95-4735-7_12

**Fig. 12.1** Large-scale afforestation in the Mu Us Sandy Land. As part of TNSDP, planting is progressing while large-scale preparation of planting sites is carried out using tractors

**Fig. 12.2** Large-scale afforestation in Hebei Province. As part of BSSC, afforestation requiring long-term management is progressing to restore farmland to forest

## 12.2 Benefits and Limitations of Afforestation

Planting trees is the most effective way to combat aeolian desertification. For example, in Jingbian County, on the southern edge of the Mu Us Sand Land, afforestation focused on preventing desertification by aeolian sand and stabilizing shifting sand dunes began in 1949. However, it was not until the aerial spraying in the TNSDP

**Fig. 12.3** Afforestation in drylands. There are two approaches: planting large seedlings with significant labour and high costs (left), or planting numerous small seedlings over a wide area (right)

launched in 1978 that the area of afforestation finally began to increase. Nevertheless, in most places, the rate of desertification was faster than vegetation recovery. For example, in the Horqin Sandy Land in the 1980s, a large-scale environmental restoration project resulted in more than 5000 $km^2$ of afforestation was countered by over twice that amount of land turning to desert in the meantime (Wu and Ci 2002). Finally, it was only in 2000 that vegetation increased in 76% of the Horqin Sandy Land, and forest area exceeded that of the 1980s (Zhou et al. 2015) (Fig. 12.3).

However, some researchers have questioned the effectiveness of mitigation programs through a series of large-scale afforestation in drylands, such as the TNSDP and NFPP (Qi et al. 2015). These programs focus on increasing the vegetation cover by prohibiting grazing, planting shrubs and trees and sowing grasses, and constructing shelter forests to protect farmland from windblown sand. In many places, there is insufficient rainfall to sustain big woody shelter forests for a long term, and supplementary irrigation is needed to ensure the survival of the trees, which depletes the limited water resources in the region. Furthermore, as greenery trees grow, evapotranspiration may increase, potentially depleting water sources. Trees competing with herbaceous plants for water could lead to reduce ground vegetation cover, thereby potentially accelerating soil erosion. Additionally, because trees grow slowly, it takes 20–30 years for the vegetation to reach a size where it can prevent sand movement effectively, and the low survival rate during this period of time makes it cost-ineffective in terms of both labor and funds. In many drylands, restoration using herbaceous vegetation produces better results than using trees or shrubs. While China's large-scale national ecosystem restoration projects, such as the TNSDP, have improved the vegetation cover in many areas, maintaining forests in drylands requires significant long-term use of a large amount of water resources, which poses a potential risk to the local water balance (Fig. 12.4). More attention should be paid to increasing the use of tree species with specific water-use characteristics, such as *Juniperus sabina* and willows discussed in this book, through updating species inventories and developing silvicultural techniques.

**Fig. 12.4** Windbreak of big poplar trees in Zhangye (Gansu Province). As the windbreak grows larger and its windbreak effect increases, the water resources available for crops decrease

## 12.3 Tackling Desertification/Land Degradation

Addressing the issue of desertification/land degradation begins by clarifying the causes and the current conditions of degradation and then determining whether the degraded land can be restored. Land degradation progresses through a dynamic and complex process, but the advancement of degradation is primarily assessed through changes in biophysical aspects. However, it should not be forgotten that human activities account for a half of the causes of desertification/land degradation and that socioeconomic factors play a strong role in whether desertified land can be restored (Fig. 12.5).

Moreover, there is no common way for immediate solution to prevent the expansion of degraded land and to restore its productive capacity, as different local natural and social conditions require different measures (Thomas 1997). Besides, the way each degrading factor operates varies according to the spatial and temporal scale (Zhuge et al. 2019). While long-term environmental changes caused by climate

**Fig. 12.5** A herder pushing cattle out of plantation area

change and rapid land degradation initiated by human activities cannot be treated separately, understanding the mechanisms of desertification and their recovery requires fundamental researches on the effect of degrading factors on the characteristics and dynamics of the dryland ecosystems, as explained so far in this book. Long-term monitoring of local ecosystems based on those accurate findings is essential to assess the importance and urgency of the challenges facing people's livelihoods and local ecosystems.

The system to tackle desertification/land degradation should recognize where and how land degradation is occurring, take timely measures in response to the land degradation process, and provide feedback on the effectiveness of the measures during implementation. The authors of this book hope to contribute to increase awareness of the need to understand the dynamics and eco-physiological characteristics of plant communities in order to take sustainable environmental restoration measures for degraded dryland ecosystems (Fig. 12.6).

**Fig. 12.6** Continuous long-term survey on the dynamics of *Juniperus sabina* community in the Mu Us Sandy Land

## References

Chen Y, Tang H (2005) Desertification in North China. Land Degrad Dev 16:367–376

Lyu Y et al (2020) Desertification control practices in China. Sustainability 12. https://doi.org/10.3390/su12083258

Thomas DS (1997) Science and the desertification debate. J Arid Environ 37:599–608

Wu B, Ci LJ (2002) Landscape change and desertification development in the Mu Us sandland, Northern China. J Arid Environ 50:429–444

Zhou et al (2015) Long-term vegetation changes in the four mega-sandy lands in Inner Mongolia, China. Landsc Ecol 30:1613–1626

Zhuge W et al (2019) Spatial-temporal pattern of human-induced land degradation in Northern China in the past 3 decades—RESTREND approach. Int J Environ Res Public Health 16. https://doi.org/10.3390/ijerph16132258

# Part VI
# Overview of Desertification Control Technologies in China

## Introduction

This part first provides an overview of China's desertification control measures, demonstrating the effectiveness of the ecological restoration project through the progression of desertification and subsequent recovery in the Mu Us Sandy Land. It also outlines the methods and effects of biological, agricultural, and engineering measures. These are ecologically and economically effective strategies for each land type, designed to mitigate the impact of wind and rain on land and vegetation, thereby preventing damage and environmental degradation. Next, from the perspective of infrastructure development, the differences in how soil degradation occurs depending on soil type, whether sandy or clayey, are described. Following a detailed description of the structure and construction methods for sand barriers using plants, clay, and gravel, and as well as for checkerboard sand barriers, the creation of windbreaks and sand-control forests is elaborated upon, along with planting methods such as direct-slip planting with long branches and aerial seeding. List the tree species used for re-afforestation on the Ordos Plateau and describe their management. Finally, based on the case of the Mu Us Sandy Land, which had become barren land but was restored to a land lush with green trees through desertification control projects, future prospects for desertification control are presented.

# Chapter 13
# History of Desertification Control Measures

**Ken Yoshikawa and Linhe Wang**

## 13.1 Desertification Control in China

Four major sandy lands and eight major deserts of China are distributed in the northwest, northeast, and north-central regions of China (the Three North Regions). China is one of countries most severely affected by desertification in the world. Since the establishment of the People's Republic of China, a series of important policy measures have been taken for the restoration and management of the ecological environment of sandy lands and deserts, such as the enactment of the "Sand Prevention and Control Law" and the formulation of the "National Sand Prevention and Control Plan." These measures launched several important restoration programs for desertification/land degradation, such as Beijing-Tianjin Sand Source Control Program, Three-North Shelterbelt Development Program, Natural Forest Protection Program, and Conversion of Cropland to Forest Program.

The Mu Us Sandy Land has large reserves of oil and natural gas and is rich in weather factors for power generation, such as strong winds and intense solar radiation. These resources provided a strong motivation for private capital to invest in the restoration, development, and management of desertified land, which led to Ordos City having a higher GDP than Shanghai City around 2011. With such a background, since 2000, large amounts of funding have been invested in the restoration of arid ecosystems in the Mu Us Sandy Land. For example, in the southern part of the Mu Us Sandy Land, Yulin City, the environment has been greatly improved through the large-scale construction and maintenance of shelter-forests, which have significantly reduced the number of windy days. As a result, agricultural and livestock production, which had once declined, is gradually increasing, and farmers and herders who had to migrate due to the progression of desertification are gradually returning. In other words, the beautiful scenery of the Mu Us Sandy Land, with its fertile water and grass and many sheep feeding in the pasture, is gradually returning.

 
K. Yoshikawa et al., *Dryland Vegetation in the Sand-Shifting Environment*,
https://doi.org/10.1007/978-981-95-4735-7_13

**Fig. 13.1** Large-scale tree planting to prevent desertification. Afforestation in drylands is approached differently depending on the degree of desertification. Even when planted in suitable sites, not all necessarily survive and thrive

There are two methods of restoring desertified lands: artificial restoration and restoration using natural forces. The restoration methods are closely related to the degree of desertification (Fig. 13.1). While the degree of desertification is moderate or mild, it is basically possible to restore the desertified land with artificial restoration alone. However, in severely desertified lands where the movement of shifting sand is active, natural resilience is also important in addition to artificial vegetation restoration.

In the Mu Us Sandy Land, the area of desertified land increased until 2000 due to people's agropastoral activities and corporate mineral resource development. However, after the implementation of ecological restoration projects as mentioned above, the progression of desertification/land degradation was suppressed, and the area of desertified land significantly decreased.

## 13.2 Biological, Agricultural, and Engineering Measures

There are three major types of countermeasures for desertification/land degradation. Biological measures include vegetation restoration through afforestation and seeding and vegetation improvement through plant community management. Agricultural measures that prevent desertification while utilizing the land for production include

**Fig. 13.2** Agricultural measure to leave high stamps on the field after harvest (Naiman, Inner Mongolia Autonomous Region)

plowing and crop rotation to promote the infiltration of surface runoff into the ground, non-tillage and intercropping to prevent the land surface from becoming bare, and high stem cutting to leave high stumps to prevent wind erosion on fields after harvest (Fig. 13.2). Engineering measures that physically prevent the movement of shifting sand include sand barrier for fixing shifting sand, water catchment systems and irrigation channels to secure water sources, and check dams to reduce the velocity of the water flow and promote mud deposition.

Depending on the soil quality, distance from water sources, and ease of plant growth in each location, different measures to combat desertification/land degradation are required, which are ecologically and economically effective for each site. Of course, it is impossible to change excessive wind speed and rainfall amounts caused by topography and climatic conditions, but the impact of wind and rain on land and vegetation can be reduced to prevent damage and environmental degradation. In regard to wind erosion, increasing the surface roughness can reduce wind speed and suppress shifting sand. Against water erosion, surface flows are dispersed so that they do not concentrate in one place, reducing the erosive power of water. When the ground surface is stabilized, a virtuous cycle begins in which vegetation cover increases as plants are more likely to establish, thereby increasing vegetation as the ground cover becomes more protected. Perennial herbaceous and woody vegetation is particularly effective, stabilizing the ground surface over the long term and preventing the movement and expansion of sand dunes.

In the Mu Us Sandy Land, a mosaic distribution of various types of sand dunes and desertified lands is found due to different levels of erosion caused by the complexity of landforms, Since the Mu Us Sandy Land is mostly sandy, the first step is to reduce wind speed and flatten the shifting sand dunes before introducing plants through seeding or afforestation. If irrigation can be achieved by digging wells, the survival rate of planted vegetation can be improved and enhance the effect of sand stabilization. Once shelter forests are established, agricultural development will also progress, further contributing to combat desertification/land degradation (Figs. 13.3 and 13.4).

**Fig. 13.3** Plum orchard intercropping with melon to prevent wind erosion (Tuolanton, Inner Mongolia Autonomous Region)

**Fig. 13.4** Paved irrigation channel (left) and small ditch within a windbreak (right) (Hetian, Xinjian Uygur Autonomous Region)

# Chapter 14
# Infrastructure Development

**Ken Yoshikawa and Linhe Wang**

## 14.1 Sandy Soil and Clayey Soil

Different soil types, such as sandy or clayey, cause soil degradation in different ways and therefore require different measures (Fig. 14.1). In sandy soils with sparse vegetation, like the Mu Us Sandy Land, wind blows the sand away and creates sand dunes. These dunes can move and sequentially destroy vegetation. If plants can be established on the sand dunes, the windbreak effect will last longer, but introducing plants in locations where the sand is moving is not easy. Therefore, before planting, sand control hedges or checkerboard sand barriers arranged in rows or grids are used to reduce wind velocity and stop sand movement in preparation for the introduction of vegetation. Alternatively, the ground surface is covered with brushwood, plastic, or chemicals in covering work to suppress the generation of shifting sand itself.

On the other hand, in clayey soils where vegetation has been lost, water erosion caused by rainwater flowing over the ground surface and collecting in depressions must be stopped. Flattening slopes of the hillside with check dams and promoting rainwater to infiltrate into the ground through water channels and ploughing reduces the volume and velocity of runoff, thus mitigating water erosion.

## 14.2 Sand Barrier

The closer to the surface, the more sand particles are blown by the wind, so increasing the surface roughness is effective to reduce the amount of flying sand due to the reduction of the wind speed at the surface. Wind and sand barriers such as sand accumulating hedges made of wooden boards or plant branches and enbankments using vegetation sandbags are effective to increase the surface roughness (Murai

 
K. Yoshikawa et al., *Dryland Vegetation in the Sand-Shifting Environment*,
https://doi.org/10.1007/978-981-95-4735-7_14

**Fig. 14.1** Water erosion in clayey soil (left: Tuolantun) and wind erosion in sandy soil (right: Ejina) (Inner Mongolia Autonomous Region)

1982) (Fig. 14.2). Without sand barriers, shifting sand would directly erode the ground surface, and dunes would move forward and backwards in response to changes in wind direction, but sand barriers can prevent these movements. Some sand barriers, such as sand accumulating hedges, increase surface roughness and stabilize and protect the ground surface by depositing and accumulating flying sand and dust around then, while others, such as sand control hedges, can provide windbreak effect without flying sand deposition. The height of the sand barriers and the wind penetration are determined by the local conditions and the shape of the barriers, such as linear or grid-like.

When multiple rows of sand barriers are constructed in a line, the wind speed changes significantly inside and outside the rows of sand barriers. In front of the sand barrier, the wind speed at a height of 2 m is the same as that of bare ground, but it slows down inside and above the sand barrier, and slows down even more behind it. On the other hand, the wind speed at a height of 0.5 m is slower than at 2 m and is relatively stable at any part of the sand barrier, showing a higher windbreak effect closer to the ground surface. For example, when a sand barrier is created with *Salix psammophila*, a clumping shrub, the ground surface roughness is highest inside the barrier and about 70% behind it, with protected distance three to four times the height of the tree. Moreover, it is 15 times higher than the roughness in the absence of sand barriers (Ffolliott 1994).

## 14.3 Checkerboard Sand Barrier

A checkerboard sand barriers is a type of sand control hedge, in which straw, tree branches, and grass stems are pressed into the sand to form a 1 m × 1 m or 1 m × 2 m grid of low fences 10–30 cm high to reduce wind speed on the dune surface and suppress sand movement. Even with barriers of several tens of cm in height, the ground surface roughness at 20 cm above ground level is 200 times higher than on the dune surface, and wind speed at 2 m above ground level is slowed by 20–40%.

**Fig. 14.2** Simple sand barrier made of willow branches

The checkerboard sand barrier is a technology developed to protect the Baolan Railway from shifting sands at Shapotou near Lanzhou, where the Tengger Desert flows into the Yellow River. The checkerboard sand barrier, 500 m and 300 m wide on the windward and leeward sides of the railway, respectively, provides tremendous windbreak effects.

Barley straw has been used as the material for the checkerboard sand barrier since the technology was developed, but corn, sunflower stalks, and short-cut willow branches have also been used instead of barley straw. The method is to line up the straw horizontally, push the middle of the straw into the sand with a shovel, and the ends of the straw naturally stand up to form a sand barrier. Construction is costly and time-consuming, but once familiar with the method, a single person can make more than 1 ha of 1-meter square grid in a day (Fig. 14.3), and the process is simple and highly effective.

The checkerboard sand barrier itself is only a temporary barrier that will weather away in a few years, and while its windbreak effect lasts, it must be planted with shrubs and herbaceous plants known as psammophytes that are resilient to drought and sand movement, called sand-binding plants. *Salix flavida, Corethrodendron scoparium (Hesysarum scoparium)*, and *Caragana korshinskii* are the most common sand-binding plants (Zhu et al. 1988), while the shrubs *Calligonum mongolicum* and *Tamarix* spp. are often used (Fig. 14.4).

**Fig. 14.3** Checkerboard sand barrier under construction (Hailar, Inner Mongolia Autonomous Region). Workers gather straw with shovels and press it into the sand

**Fig. 14.4** Several types of checkerboard sand barriers made of (**a**) straw, (**b**) sunflower stalks, (**c**) net woven barley culms, (**d**) short-cut willow branches

## 14.4 Clay Sand Barrier

If rainfall is sufficient for plant survival, sand dune stabilization through planting as mentioned above is possible, but where rainfall is low and plant growth is difficult, physico-chemical sand control measures are required.

When clay is mixed with water, it becomes highly adhesive due to the high affinity between clay particles and water, and the clay is no longer a single free soil particle but a soil mass with a specific structure. Once the clayey soil mass is formed, its structure is difficult to destroy. Therefore, clay can be stacked in lines or grids on the upper windward slope of the sand dune to be used as a barrier. In such case, the height of the embankment is 30–40 cm, with a spacing of 1.5–2 m.

The clay is sprayed on the dune surface or applied with a trowel. Heavy clay is most suitable for spraying. Water is added to the clay to form a watery paste, and sometimes emulsion is added to enhance adhesion. The adjusted clay is evenly sprayed over the surface of the sand dune, but the range, amount, and thickness of the spray vary depending on the degree of slope of the sand dune. When applying, the clay is made into a paste and spread evenly on the surface of the sand dune using a trowel. The thickness is generally 3–5 cm but depends on the degree of wind erosion of the sand dune. In both cases, work is done from the top to the bottom of the sand dune to avoid trampling by the workers.

## 14.5 Stabilization of Sand Dune Surfaces with Chemicals

In some cases, instead of clay, diluted chemicals that can solidify sand, such as polyvinyl alcohol, are sprayed onto the surface of heavily wind-eroded shifting sand dune. The moisture of the sprayed chemical penetrates into the lower layers, but the chemical binding substance remains in the gaps of the sand layer about 1–5 mm thick, wrapping each sand particle with a protective film and binding them together. This protective film prevents wind and water flow from coming into direct contact with the shifting sand surface, thereby preventing erosion. Some dunes are coated with waste oil or asphalt emulsion, as is done in the Middle East. These methods can only secure the sand in place, but they do not prevent the sand particles from being blown away by the wind.

## 14.6 Sand Barriers with Gravel

There is also a method of stacking gravel to create a grid-like sand barrier. Gravel is sturdy, does not break down with wind erosion or crumble in rain, and has high thermal insulation properties. Therefore, when gravel is spread across the surface of a sand dune (stone mulch), it can reduce the occurrence of shifting sand and also suppress moisture evaporation from the ground. In addition, moisture can be trapped through condensation caused by the insulating effect of gravels.

## References

Ffolliott PF (ed) (1994) Dryland forestry – planning and management. Wiley, p 453

Murai H (1982) Greening situation for soil and water conservation in the desert and loess zone of the Yellow River basin. Green Technol 8(3):15–28 (in Japanese with English summary)

Zhu Z, Liu S, Di X (1988) Desertification and rehabilitation in China. The International Center for Education and Research on Desertification Control, Lanzhou, p 222 (in Chinese)

# Chapter 15
# Introduction of Plants

**Ken Yoshikawa and Linhe Wang**

## 15.1 Aerial Seeding

Aerial seeding using airplanes is an effective method for large-scale greening because it is quick, low-cost, and requires little labor (Fig. 15.1). However, the effectiveness depends largely on weather conditions and location (Wei et al. 2000). Aerial seeding in the Mu Us Sandy Land began in the late 1950s, and large-scale seeding projects were implemented from 1983 to 1995, during which time 1000 km$^2$ of Mu Us Sandy Land and 200 km$^2$ of the Kubuqi Desert were greened.

After the seeds are sown from the air, they must quickly germinate and establish themselves at the site where the sand is moving violently. Therefore, the plant seeds used for aerial seeding must have high germinative capacity and rapid rooting characteristics to withstand various environmental stresses such as drought, poor nutrition, and wind burial (Forestry and Sand Control Research Institute of Yeke Juu League 1991).

In the Mu Us Sandy Land and the Kubuqi Desert, native shrubs and herbs are often used, such as *Corethrodendron lignosum (=Hedysarum mongolicum), Artemisia sphaerocephala, Astragalus laxmannii (=Astragalus adsurgens), Caragana microphylla, Amorpha fruticosa, Melilotus officinalis (=Melilotus suavealens), Agriophyllum squarrosum,* and *C. scoparium*. There are also cases where tall trees such as pines and *Ulmus pumila* are used.

Light seeds are blown away by the wind when aerially sprayed without treatment and the desired area cannot be revegetated. Therefore, when sowing light seeds, the seeds are coated with fine sand or clay to increase their weight. In addition, fertilizers or water retention agents may be mixed into the coating to promote germination and establishment. For relatively large seeds, such as pines, which are easily eaten by rodents after sowing, repellents may be applied.

In the Mu Us Sandy Land, when three shrub species, *A. sphaerocephala, Corethrodendron lignosum* var. *leave (=Hedysarum leave)*, and *C. scoparium*, are

 
K. Yoshikawa et al., *Dryland Vegetation in the Sand-Shifting Environment*,
https://doi.org/10.1007/978-981-95-4735-7_15

**Fig. 15.1** Aircraft landing on grassland for aerial seeding

sown, each species becomes established and forms community. Subsequently, without any human intervention, *Artemisia ordosica* invades and forms mixed communities with *Haloxylon leave* or *Haloxylon scoparium*, eventually becoming a pure community of *A. ordosica*. This is because the sown seeds formed communities, which fixed the sand and reduced soil moisture, resulting in poor growth of each community. Since *H. scoparium* and *H. leave* have higher economic value than *A. ordosica*, management is also necessary to prevent the progression of this vegetation succession (Desertification Prevention Group 2015).

## 15.2 Windbreaks and Sand Control Forests

The wind and sand control functions of large trees are manifested in three aspects: (1) covering the ground surface to protect from wind action, (2) reducing wind speed up to a certain height from the ground surface and prevent wind erosion at the ground surface, and (3) blocking the movement of flying sand particles to promote their deposition, and further, the roots can bind sand particles and stabilize sandy ground (Fig. 15.2). In addition, matter production by photosynthesis improves the soil fertility by supplying litter (fallen leaves and dead branches) to the soil. The growth of artificially restored vegetation is faster than the natural restoration rate, so artificial restoration works aimed at vegetation recovery can improve the soil nutrient status more quickly.

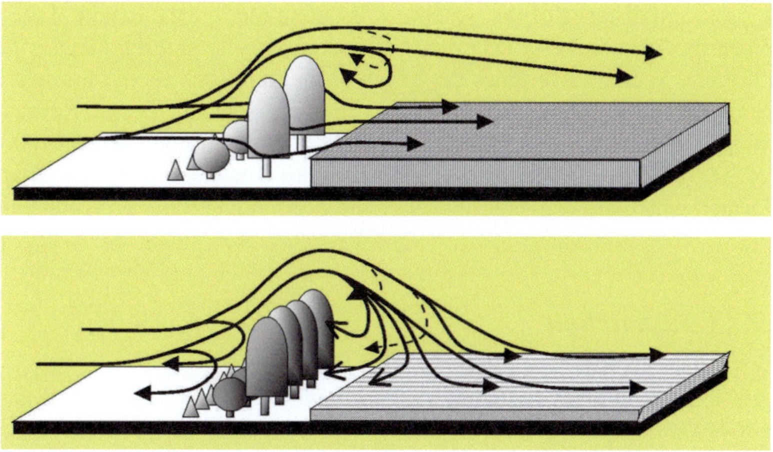

**Fig. 15.2** Effect of windbreak on wind flow. Sparse windbreak (top) allows wind to pass through, extending its windbreak effect over a greater distance. Dense windbreak (bottom) causes turbulence to form on the leeward side, limiting the effective range of its windbreak effect

## *15.2.1 Structure*

In drylands with strong winds and no natural barriers such as topographical undulations, windbreaks are an effective way to protect houses and farmlands from the wind. Windbreaks usually consist of trees or shrubs planted in one or several rows, which reduce wind speed and promote crop growth in farmland. However, large trees consume significant amounts of water to survive, so there must be sufficient rainfall or groundwater.

The annual rainfall in the Mu Us Sandy Land is only about 360 mm, but the abundance of groundwater makes it possible to maintain a considerable size of tree belt. For this reason, most farmlands are protected by windbreaks that are perpendicular to the strong northwest seasonal winds blowing in spring, which carry yellow sand to Japan. In places where the wind direction is not constant, windbreaks are built in a grid pattern, and in places with undulating terrain, it should be built along the top of the convex terrain while avoiding slopes.

The windbreak effect is determined by the porosity and height of the forest belt. Porosity is the proportion of the forest belt that is not obstructed by trunks, branches, and leaves when viewed from the front. If the porosity is too low, the wind slows down too much in front of the forest belt, leading to the accumulation of sand. Furthermore, wind blowing over the forest belt swirls on the leeward side, which gouges the ground surface and increases the damage from flying sand. The optimal porosity for the highest windbreak effect is around 40%, and it is better if the gaps are small and evenly dispersed over the forest surface (Torita and Fukuchi 1997). Therefore, windbreaks are more effective if trees are planted at intervals that allow some wind to pass through.

The protected distance of the windbreak is determined by the height of the forest belt. When the wind strikes a windbreak of optimal density at a right angle, the protected distance is about five times the height of the forest belt on the windward side and about 20 times on the leeward side (The Japanese Society of Revegetation Technology 2005). For a 10 m high forest belt, the next forest belt is created at a distance of 200 m. Although a higher forest belt is more efficient, it uses a larger amount of soil water, which has a greater impact on the surrounding water environment.

### *15.2.2 Construction*

The tree species suitable for windbreaks should be wind-resistant, like poplars, which do not bend their trunks even when the strong wind blows and spread a dense, uniform canopy. Fast-growing tree species are desirable to realize windbreak effects as quickly as possible. Also, to avoid competition with crops for water, it is preferable that their root systems do not spread horizontally. Since no single species can fulfil all these characteristics, combining several species can create a forest belt with high windbreak effects.

Since wind erosion is the greatest constraint to afforestation on shifting sand dunes, forest belts cannot be created without preventing the sand movement in the afforestation site. Therefore, wind barriers with fast-growing shrubs or checkerboards sand barrier, which are suited to the site conditions, should first be implemented to secure planting site. After the wind weakens and the sand stops moving, full-scale tree planting should be undertaken (Fig. 15.3).

**Fig. 15.3** Windbreaks combining trees and shrubs. (**a**) Start with shrubs, (**b**) introduce trees after shrubs have established, (**c**) remove shrubs, and (**d**) thinning trees

## 15.3 Forestland

By selecting the right species for the right site, adapting to the local moisture condition, as in the case of the Three-North Shelterbelt Development Program, it is possible to create forestland with a wider area and more extensive spread than rows of windbreaks (Fig. 15.4).

### *15.3.1 Tree Species*

As many shrub species grow on the Ordos Plateau, many native species with high adaptability to desertification/land degradation can be used to create forestlands. For example, *Haloxylon ammodendron, Calligonum mongolicum, Tamarix* spp., *C. microphylla, Artemisia* spp., *Salix* spp. are suitable shrub species. Tamarisk is a phreatophyte that can use deep groundwater by extending its taproot deep into the ground.

Also, for the greening of the Mu Us Sandy Land, shrub species such as *Juniperus sabina (=Sabina vulgaris), A. ordosica, C. korshinskii, Salix psammophila, Salix gordejevii,* and *C. lignosum* var. *leave*, as well as tall tree species like *Pinus tabuliformis, Ulmus pumila,* and *Populus* spp. can be used. *J. sabina* is also excellent in

**Fig. 15.4** Artificial forests have a greater spatial extent than linear windbreaks (Hailar, Inner Mongolia Autonomous Region)

cold and drought resistance, forms stable communities for several decades, and has excellent ecological functions such as soil formation. Additionally, other economically useful tree species such as *Hippophae rhamnoides, Salix matsudana,* and *Ephedra sinica* are widely planted, and the exotic species *Pinus sylvestris* var. *mongolica* is planted as a useful timber species.

### *15.3.2 Planting*

Direct-slip planting with long branches is one of the traditional ways for planting *S. matsudana*. Large branches of *S. matsudana*, about 4 cm in diameter (3–4 years old), are cut from scion stools; all twigs are removed, and the branch is made into stick-like large cutting about 3 m long. Then, about 1 m of the base of the large cuttings is buried in sand, leaving about 2 m of a stake-like branch above ground. When planted in favorable moisture conditions around May, they root within the same year and begin to grow (Kobayashi et al. 1989) (Fig. 15.5).

The appropriate planting density varies depending on the physiological and ecological characteristics of the tree species and the site conditions of the planting area. In arid areas where the annual average rainfall is less than 250 mm, shrubs such as *Haloxylon ammodendron, C. korshinskii, Calligonum mongolicum, Elaeagnus angustifolia,* and *Tamarix* spp. can be planted at a density of 200–500 trees/ha. In semi-arid areas with an annual rainfall of 250–400 mm, the appropriate planting density for *C. korshinskii* is 600–900 trees/ha, for *C. scoparium* and *H. leave* it is 800–1200 trees/ha, for *S. psammophila* it is 1000–1500 trees/ha, and for *P. sylvestris* var. *mongolica, Ulmus pumila,* and *Salix* spp. it is 200–500 trees/ha. These planting densities are in principle, and in areas with abundant groundwater like interdune depressions of the Mu Us Sandy Land, planting can be done at a higher density (Berrhahmouni et al. 2015).

**Fig. 15.5** Direct-slip planting with long willow cuttings. Immediately after inserting long willow cuttings into sandy soil (left) and subsequent growth of cuttings (right). Many branches for cuttings have developed on the top of main trunks. The leaves are fodder for livestock

### 15.3.3 Management

Some dryland forestations are maintained by rainwater alone, while others require irrigation. Even when irrigating, there are cases where irrigation continues for a long time after planting, like windbreaks around farmland, and cases where irrigation is only done immediately after planting.

If irrigation is not done at all, planting should begin after the start of the rainy season. Once the soil becomes moist up to about 20 cm deep, the seedlings can be considered to have retained enough moisture to grow for several months, and planting can begin. If planting is started before sufficient moisture has accumulated in the soil, the planted seedlings will die if the rains do not continue. On the other hand, if the start of planting is delayed, the rainy season may end before the seedlings are established, leading to the soil drying out and the seedlings withering.

To efficiently use rainwater, some devices collecting rainwater in one place are required. Various forms of microcatchments have been used. Some dig down slightly around planted seedlings, while others create V-shaped trenches or ridges to gather surface runoff at the base of the trees, which can either store water at the base of the seedlings or allow it to permeate underground.

By making the planting holes sufficiently deep and watering them adequately immediately after planting, the survival rate of the seedlings can be increased. In particular, watering during the first growing season after planting significantly increases the survival rate of seedlings. Similarly, watering in the spring of the second year is also effective. Ideally, watering should continue until the roots have developed to a sufficient depth, but in dryland afforestation where forests are being created on vast lands with minimal expenditure, it is not practical to continue watering for months after planting. In normal afforestation, it is realistic to water sufficiently just after planting and continue watering once a year for several years.

### 15.3.4 Stabilization of Shifting Sand Dunes by Planting Trees

There is a method of stabilizing shifting sand dunes by afforestation, which combines a small-scale windbreak and a sand control forest, called "Block in the front and pull from the back."

A windbreak belt of tall trees is created in the interdune depressions on the leeward side of the dune to slow down the wind blowing down the dune. In the Mu Us Sandy Land, *Salix matsudana* and *Populus* spp. are planted. At the same time, short shrubs and perennial herbs are planted on the foot of the windward side of the dune to cut off the supply of sand to the dune. In the Mu Us Sandy Land, *S. psammophila* and *A. sphaerocephala* are used.

The planting of shrubs is initially limited to the lower third of the windward slope, so that sand is carried away by strong winds from the upper part of the dune. However, since the supply of sand from the rear windward side is suppressed, the

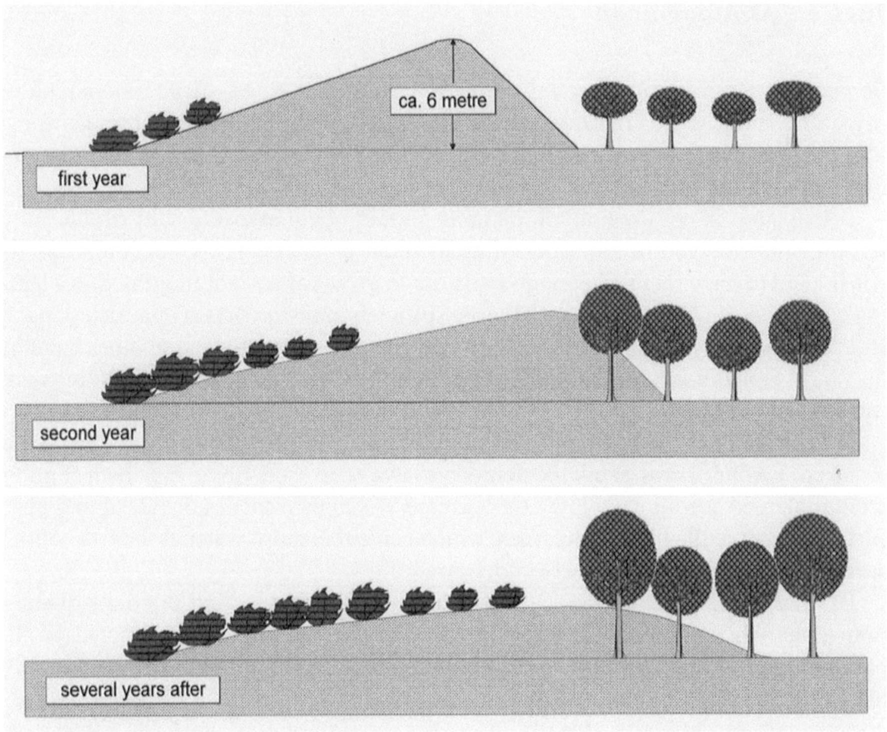


**Fig. 15.6** Schematic diagram of the "Block in the front and pull from the back" method for stabilizing shifting sand dunes

height of the sand dune gradually decreases. At the same time, the sand carried away by the wind is trapped in the tall forest belt on the leeward side, so the dune becomes flatter. The following year, similar planting is done on the windward side slightly above the part planted the previous year. As a result, the shifting sand dunes are gradually eliminated and transform into highly productive pastures (Yao 1986) (Fig. 15.6).

## References

Berrhahmouni N, Regato P, Parfondry M (2015) Global guidelines for the restoration of degraded forests and landscapes in drylands. FAO, p 148

Desertification Prevention Group (2015) Sustainable environmental conservation in desertification advancing regions of East Asia. Hanashoin, p 201 (in Japanese)

Forestry and Sand Control Research Institute of Yeke Jun League (1991) Technical report on the application and promotion of airplane seeding afforestation and sand control technology in Yikezhao league, Inner Mongolia. p 18 (in Chinese)

Kobayashi T et al (1989) Relationship between site and growth of large branch direct planting of Populus in Mu Us sandy land. J Jpn Soc of Reveget Technol 15:1–8 (in Japanese with English summary)

The Japanese Society of Revegetation Technology (ed) (2005) Encyclopedia of environmental greening. Asakura Publishing, p 484 (in Japanese)

Torita H, Fukuchi M (1997) Function of windbreak forest -2 research on thinning of windbreak forest considering windbreak effect. Hokkaido For Exp Station Rep 34:1–6 (in Japanese)

Wei J et al (2000) Factors affecting the survival area rate of greening plants by airplane seeding – targeting the Yeke Jun league region in the Inner Mongolia Autonomous Region. Agri Soil Collect 200:105–111 (in Japanese with English summary)

Yao H (1986) On greening technology of desert. Green Technol 12(1):29–43 (in Japanese with English summary)

# Chapter 16
# Prospects for Greening Technology

**Ken Yoshikawa**

"The wind blows the yellow dust, making it hard to open one's eyes. Crops fail to grow, houses are buried, and people are leaving. When the yellow dust comes, everyone shakes their heads in despair." This song accurately depicted the harsh environment of the Mu Us Sandy Land in the past. As mentioned above, the Mu Us Sandy Land was not barren from the beginning; a thousand years ago, it was rich in water and grass, with many cattle and sheep. However, changes in the natural environment, along with the expansion of human activities such as illegal logging, unplanned reclamation, and overgrazing, have led to the gradual desertification/land degradation of the Mu Us Sandy Land, transforming vast green fields into a massive sea of sand.

However, the desertified land area has been reduced in all three sandy lands, including the Mu Us Sandy Land, where many subsequent large-scale desertification control projects have been implemented. As a result, the number of days with sandstorms in a year has decreased from 6.8 to 2.4 days in the Mu Us Sandy Land, and from 5.1 days to less than 1 day in the Beijing-Tianjin-Hebei regions. These successes were not only due to large-scale projects but also due to individual measures taken by farmers and herders, such as letting their farmland lie fallow in rotation, reducing the number of livestock to conserve the productivity of pastures, which greatly restored the desertified land around their residential area.

The forest coverage rate in Yulin City has increased from 0.9% at the time of the founding of China to 33%, and all of the approximately 6000 km$^2$ of shifting sand land has been fixed. The former "Kingdom of Shifting Sand" has been transformed into "The dryland north of the Great Wall, but as fertile and beautiful as the southern Changjiang River" where green trees thrive. Ordos City has been managing the Mu Us Sandy Land for over 60 years and has achieved a historical transition from "desertification progresses and people disappear" to "desertified land retreats and people return." The key to successful ecosystem management is to continue long-term and consistent efforts (Fig. 16.1).



K. Yoshikawa et al., *Dryland Vegetation in the Sand-Shifting Environment*,
https://doi.org/10.1007/978-981-95-4735-7_16

**Fig. 16.1** Greening in dryland

MIX
Papier aus verantwortungsvollen Quellen
Paper from responsible sources
FSC® C105338

If you have any concerns about our products,
you can contact us on
**ProductSafety@springernature.com**

In case Publisher is established outside the EU,
the EU authorized representative is:
**Springer Nature Customer Service Center GmbH**
**Europaplatz 3, 69115 Heidelberg, Germany**

Printed by Libri Plureos GmbH
in Hamburg, Germany